LIGHTING
OUT
FOR THE
TERRITORY

Lighting Out

REFLECTIONS
ON
MARK TWAIN
AND
AMERICAN
CULTURE

for the Territory

SHELLEY FISHER FISHKIN

New York   Oxford

OXFORD UNIVERSITY PRESS

1997

OXFORD UNIVERSITY PRESS

Oxford New York

Athens, Auckland, Bangkok, Bogota, Bombay,
Buenos Aires, Calcutta, Cape Town, Dar es Salaam, Delhi,
Florence, Hong Kong, Istanbul, Karachi,
Kuala Lumpur, Madras, Madrid, Melbourne,
Mexico City, Nairobi, Paris, Singapore
Taipei, Tokyo, Toronto
and associated companies in
Berlin, Ibadan

Published by
Oxford University Press, Inc.,
198 Madison Avenue, New York,
New York 10016

Oxford is a registered trademark of
Oxford University Press

Library of Congress
Cataloging-in-Publication Data
Fishkin, Shelley Fisher.
  Lighting out for the territory : reflections on Mark
Twain and American culture / by Shelley Fisher
Fishkin.
    p. cm.
  Includes index.
  ISBN 0-19-510531-1 (hardcover)
    1. Twain, Mark. 1835–1910—Knowledge—America.
  2. Literature and society—United States—History—19th
  century.   3. Twain, Mark, 1835–1910—Appreciation—
  United States.   4. National characteristics, American,
  in literature.   5. Humorous stories, American—
  History and criticism.   6. Authors, American—19th
  century—Biography.   7. United States—In literature.
  8. America—In literature.   I. Title.
  PS1342.A54F57   1996
  818'.409—dc20   96-34612
                                                              CIP

ISBN 0-19-510531-1

1 3 5 7 9 8 6 4 2

Printed in the United States of America on acid-free
paper

Prologue and Epilogue: Photograph reproduced with the permission of The Mark Twain House, Hartford, Connecticut.

Half-Title, Title, and Chapter 1: Photograph reproduced with the permission of The Mark Twain House, Hartford, Connecticut.

Chapter 2: Photograph reproduced with the permission of The Mark Twain House, Hartford, Connecticut.

Chapter 3: Photograph reproduced from the Mark Twain Collection of The James S. Copley Library, La Jolla, California.

For my father,

MILTON FISHER,

whose sense of humor

and passion for justice

prepared me to appreciate

Mark Twain

CONTENTS

Prologue, *3*

1   The Matter of Hannibal, *13*

2   Excavations, *71*

3   Ripples and Reverberations, *127*

Epilogue, *183*

Notes and Sources, *205*

Acknowledgments, *250*

Index, *255*

LIGHTING
OUT
FOR THE
TERRITORY

PROLOGUE

My mother startled me out of a cocoon of cartoons and cocoa one blustery Saturday morning when I was eleven.

"Get dressed. We're going on a mystery trip."

"But it's cold," I protested. "It's supposed to snow."

"Then dress warmly. Hurry. We've got a two-hour drive."

The snow that was beginning to transform the Connecticut landscape into a dreamscape of silver and white only increased my mother's determination to get us where we were going and back before the storm really hit. Once in the car, I badgered her until she revealed our destination: a house in Hartford, Connecticut, that had once belonged to a writer named Mark Twain and had recently been restored and opened as a museum. Why anyone would want to drive four hours in the snow just to look at a house was beyond me, but I kept that thought to myself. My mother always had her reasons, and they usually turned out to be good ones.

"We're here."

I must have dozed off because it seemed like we'd just started. I opened my eyes and looked around for a house. I wasn't prepared for the vast, ornate, orange and brown structure that stared down at us from a hilltop. And I had never imagined a home with a Tiffany-designed interior, mosaic tiles, oriental carpets, stenciled wall coverings, a banister of deep, rich, polished walnut, and a phone booth. A phone booth? Had I heard the guide right? Indeed I had. And it was complete with a report card Mark Twain used to grade the phone company. (One plus sign signified "artillery can be heard" in the phone lines, two "thunder can be heard.")

As the tour proceeded, I saw the speaking tubes in the nursery (which Twain used to convey important instructions from "Santa" every Christmas), the first flush toilet in Connecticut (at least they didn't know of any earlier ones), a fireplace with a divided flue that let you bask in the warmth of a crackling fire while watching the snowflakes fall right above it, and in the basement a strange, complicated machine that was designed to set type automatically but never quite did what it was supposed to and somehow ate all of Twain's money instead.

That whole magical afternoon my mother beamed. She knew what she was doing all right. She was planting the seeds of a lifelong fascination with the man who had lived in that house.

Soon after we returned from Hartford, she began reading *Tom Sawyer* to me as a bedtime story. I thought Huck and Tom could be a lot of fun, but I dismissed Becky Thatcher as a bore. When I was twelve I invested a nickel at a local garage sale in a book that contained short pieces by Twain. That was where I met Twain's Eve. Now *that's* more like it, I decided, pleased to encounter a female character I could identify *with* instead of against. Eve had spunk. Even if she got a lot wrong, you had to give her credit for trying. "The Man That Corrupted Hadleyburg" left me giddy with satisfaction: none of my adolescent reveries of getting even with my enemies were half as neat as the plot of the man who got back at that town. "How I Edited an Agricultural Paper Once" set me off in spasms of giggles.

People sometimes told me that I looked like Huck Finn. "It's the freckles," they'd explain—not explaining anything at all. I didn't read *Huckleberry Finn* until my junior year in high school, when it was assigned in my English class. It was the fall of 1965. I was living in a small town in Connecticut. I expected a sequel to *Tom Sawyer*. So when the teacher handed out the books and announced our assignment, my jaw dropped: "Write a paper on how Mark Twain used irony to attack racism in *Huckleberry Finn*."

A year before, the bodies of three young men who had gone to Mississippi to help blacks register to vote—James Chaney, Andrew Goodman, and Michael Schwerner—had been found in a shallow grave; a group of white segregationists (the county sheriff among them) had been arrested in connection with the murders. America's inner cities were simmering with pent-up rage that had started exploding in the summer of 1965, when riots in the Watts section of Los Angeles left thirty-four people dead. None of this made any sense to me. I was confused, angry, certain that there was something missing from the news stories I read each day: the why. Then I met Pap Finn. And the Phelpses.

Pap Finn, Huck tells us, "had been drunk over in town" and "was just all mud." He erupts into a drunken tirade about "a free nigger . . . from Ohio; a mulatter, most as white as a white man," with "the whitest shirt on you ever see, too, and the shiniest hat; and there ain't a man in town that's got as fine clothes as what he had."

> . . . they said he was a p'fessor in a college, and could talk all kinds of languages, and knowed everything. And that ain't the wust. They said he could *vote,* when he was at home. Well, that let me out. Thinks I, what is the country a-coming to? It was 'lection day, and I was just about to go and vote, myself, if I warn't too drunk to get there, but when they told me there was a State in this country where they'd let that nigger vote, I drawed out. I says I'll never vote agin. Them's the very words I said. . . . And to see the cool way of that nigger—why, he wouldn't a give me the road if I hadn't shoved him out o' the way.

Later on in the novel, when the runaway slave Jim gives up his freedom to nurse a wounded Tom Sawyer, a white doctor testifies to the stunning altruism of his actions. The Phelpses and their neighbors—all fine, up-standing, well-meaning, churchgoing folk—agreed that

> Jim had acted very well, and was deserving to have some notice took of it, and reward. So every one of them promised, right out and hearty, that they wouldn't curse him no more.
> Then they come out and locked him up. I hoped they was going to say he could have one or two of the chains took off, because they was rotten heavy, or could have meat and greens with his bread and water, but they didn't think of it.

Why did the behavior of these people tell me more about why Watts burned than anything I had read in the daily paper? And why did a drunk Pap Finn railing against a black college professor from Ohio whose vote was as good as his own tell me more about white anxiety over black political power than anything I had seen on the evening news?

Mark Twain knew that there was nothing, absolutely *nothing*, a black man could do—including selflessly sacrificing his freedom, the only thing of value he had—that would make white society see beyond the color of his skin. And Twain knew that depicting racists with chilling accuracy could expose the viciousness of their worldview like nothing else could. It was an insight echoed some eighty years after Twain penned Pap Finn's rantings about the black professor, when Malcolm X famously asked, "Do you know what white racists call black Ph.D.'s?" and answered, "*Nigger!*"

Mark Twain taught me things I needed to know. He taught me to understand the raw racism that lay behind what I saw on the evening news. He taught me that the most well-meaning people can be hurtful and myopic. He taught me to recognize the supreme irony of a country founded on freedom that continued to deny freedom to so many of its citizens. He also taught me how powerful irony and satire could be in the service of truth. I found it exhilarating to analyze why it was so important that Twain never let Huck figure out he was doing the right thing all along—why a naive narrator could be such an effective vehicle for conveying to readers the moral bankruptcy of the world in which he lived. It was exciting to read between the lines—as well as under, around, and behind them—to try to figure out what the author, as opposed to his characters, was really trying to do and how he did it.

*   *   *

In the years since I first entered his house in Hartford that winter afternoon, Mark Twain has certainly led me places neither my mother nor I could have predicted. My mother passed away twenty years ago, and one of the great sadnesses I carry with me is that she isn't here to share the adventures that she so subtly set in motion. Mark Twain has led me into the past, prompting me to examine the history of American journalism, literature, literary criticism, education, race relations, folklore, dialect, rhetorical traditions, historiography, and law; to study nineteenth-century newspapers; to interview descendants of slaves; and to ask troubling questions about why some aspects of the past are commemorated while others are ignored. Twain has led me into the future, compelling me to become conversant with new technologies, to use them to answer questions I never dreamed of asking until last week. He has led me into the present, introducing me to writers, scholars, archivists, activists, inventors, and artists. He has led me to Hannibal and Elmira and (many, many times) back to Hartford.

* * *

He has been called the American Cervantes, our Homer, our Tolstoy, our Shakespeare, our Rabelais. Ernest Hemingway maintained that "all modern American literature comes from one book by Mark Twain called *Huckleberry Finn*." President Franklin Delano Roosevelt got the phrase "New Deal" from *A Connecticut Yankee in King Arthur's Court*. *The Gilded Age* gave an entire era its name. "The future historian of America," wrote George Bernard Shaw to Samuel Clemens, "will find your works as indispensable to him as a French historian finds the political tracts of Voltaire."

Mark Twain has indelibly shaped our view of who and what the United States is as a nation and who and what we might become. He helped define the rhythms of our prose and the contours of our moral map. He saw our best and our worst, our extravagant promise and our stunning failures, our comic foibles and our tragic flaws. He understood better than we did ourselves our dreams and aspirations, our potential for greatness and our potential for disaster. His fictions brilliantly illuminated the world in which he lived and the world we inherited, changing it—and us—in the process. He knew that our feet often danced to tunes that had some-how remained beyond our hearing; with perfect pitch he played them back to us.

"In a century we have produced two hundred and twenty thousand books," Mark Twain wrote in 1906. "Not a bathtub-full of them are still alive and marketable." Yet in 1906, as well as every year since, "alive and marketable" works by Twain alone would have filled the average bathtub to overflowing. By 1906 some 90 different editions of his books had appeared in print in the United States, and hundreds had appeared abroad. By 1976 his writings in book form had been published in at least 5,344 editions in fifty-five countries and translated into seventy-two foreign languages. The trend shows no signs of abating: indeed, a recent Japanese translation by Professor Hiroshi Okubo of "Eve's Diary" and "Extracts from Adam's Diary" sold 230,000 copies—enough to fill every bathtub on Twain's block in Hartford.

Twain neither held nor sought an elective or appointed office, yet he was seen as representing his nation when he traveled abroad. Throughout the world he is viewed as the most distinctively American of American authors—and also as one of the most universal. He has been a major influence on writers in the twentieth century, from Argentina to Nigeria to the Czech Republic. Children ride horses daily at the Tom Sawyer *Bokuju* (farm or meadow), an amusement park in Japan. *Adventures of Huckleberry Finn* is taught every year in classrooms in Belfast, Buenos Aires, Lublin, Beijing, Jerusalem, Thessaloníki, Rio de Janeiro, Delhi, Tokyo, and Riyadh. "Are you an American?" Twain once jotted in his notebook: "No, I am not *an* American. I am *the* American."

A century after the American Revolution sent shock waves throughout Europe, it took Mark Twain to explain to Europeans and to his countrymen alike what that revolution had wrought. He probed the significance of this new land and its citizens and identified what it was about the Old World that America abolished and rejected. The founding fathers had thought through the political dimensions of making a new society; Twain took on the challenge of interpreting the social and cultural life of the United States for those outside its borders as well as for those who were living the changes he discerned.

Americans may have constructed a new society in the eighteenth century, but they articulated what they had done in voices that were largely interchangeable with those of Englishmen until well into the nineteenth century. Mark Twain became the voice of the new land, the leading translator of who and what the "American" was and, to a large extent, still is. Frances Trollope's *Domestic Manners of the Americans,* a bestseller in England, J. Hector St. John de Crèvecoeur's *Letters from an American Farmer*, and Tocqueville's *Democracy in America* all tried to explain America to Europeans. But Twain did more than that: he allowed European readers to *experience* this strange "new world." And he gave his countrymen the tools to do two things they had not quite had the confidence to do before. He helped them stand before the cultural icons of the Old World unembarrassed, unashamed of America's lack of palaces and shrines, proud of its brash practicality and bold inventiveness, unafraid to reject European models of "civilization" as tainted or corrupt. And he also helped them recognize their own insularity, boorishness, arrogance or ignorance and laugh at it—the first step toward transcending it and becoming more "civilized" in the best European sense of the word.

Twain understood the potential of art in the service of truth, and he grasped the potential of humor in the service of morality. ("Against the assault of Laughter," he wrote, "nothing can stand.") His unerring sense of the right word and not its second cousin taught people to pay attention when he spoke, whether in person or in print. He said things that were smart and things that were wise, and he said them incomparably well. Twain often strikes us as more a creature of our time than of his. He appreciated the importance and complexity of mass tourism and public relations, fields that would come into their own in the twentieth century but were only fledgling enterprises in the nineteenth. He explored the liberating potential of humor and the dynamics of friendship, parenting, and marriage. He narrowed the gap between "popular" and "high" culture, and he meditated on the enigmas of personal and national identity. Indeed, it would be difficult to find an issue on the horizon today that Twain did not touch on somewhere in his work. Heredity versus environment? Animal rights? The boundaries of gender? The place of black voices in the cultural heritage of the United States? Twain was there.

With startling prescience and characteristic grace and wit, Twain zeroed in on many of the key challenges—political, social, and technological—that would face his country and the world for the next hundred years: the challenge of race relations in a society founded on *both* chattel slavery and ideals of equality, and the intractable problem of racism in American life; the possibilities of new technologies to transform our lives in ways that could be both exhilarating and terrifying—as well as unpredictable; the problem of imperialism and the difficulties entailed in getting rid of it. But he never lost sight of the most basic challenge of all: each man or woman's struggle for integrity in the face of the seductions of power, status, and material things.

Samuel Clemens entered the world and left it with Halley's Comet, little dreaming that generations hence Halley's Comet would be less famous than Mark Twain. There is a Mark Twain Bank in St. Louis; a Mark Twain Diner in Jackson Heights, New York; a Mark Twain Smoke Shop in Lakeland, Florida; and an Asteroid Mark Twain in outer space. Neatly reflecting, perhaps, the spectrum of interpretations of Twain's legacy, there is a Mark Twain Junior High for emotionally disturbed youngsters in Rockville, Maryland, and a Mark Twain High School for the Gifted and Talented in Brooklyn, New York. Mark Twain's image peers out at us from advertisements for Bass Ale and Old Crow Bourbon (his drink of choice was scotch), as well as for a gas company in Tennessee, a hotel in the nation's capital, and a cemetery in California.

Ubiquitous though his name and image may be, Mark Twain is in no danger of becoming a petrified icon. On the contrary, Mark Twain lives. *Huckleberry Finn* is "the most taught novel, most taught long work, and most taught piece of American literature" in American schools from junior high to graduate school. Hundreds of Twain impersonators appear in theaters, trade shows, and shopping centers in every region of the country. Scholars publish hundreds of articles as well as books about Twain every year, and he is the subject of daily exchanges on the Internet. A journalist somewhere in the world finds a reason to quote Twain just about every day. Television series such as *Bonanza, Star Trek: The Next Generation*, and *Cheers* broadcast episodes that feature Mark Twain as a character. Hollywood screenwriters regularly produce movies inspired by his works, and writers of mysteries and science fiction continue to weave him into their plots.

Mark Twain entered the public eye at a time when many of his countrymen considered "American culture" an oxymoron; he died four years before a world conflagration that would lead many to question whether the contradiction in terms was not "European civilization" instead. In between he worked in journalism, printing, steamboating, mining, lecturing, publishing, and editing, in virtually every region of the country. He tried his hand at humorous sketches, social satire, historical novels, children's books, poetry, drama, science fiction, mysteries, romances, philosophy, travelogues, memoirs, polemics, and several genres no one had ever seen before or has ever seen since. He invented a self-pasting scrapbook, a history game, a vest strap, and a gizmo for keeping bedsheets tucked in; he put money into machines and processes designed to revolutionize typesetting and engraving; and he invested in a food supplement called "Plasmon." Along the way he cheerfully impersonated himself and prior versions of himself for doting publics on five continents, while playing out a charming rags-to-riches story followed by a devastating riches-to-rags story followed by yet another great American comeback. He had a long-running real-life engagement first in a sumptuous comedy of manners and then in a tragedy not of his own design: during the last fourteen years of his life almost everyone he ever loved was taken from him by disease and death. How can we come to know this larger-than-life figure who managed to leave his mark on so many aspects of his world and our own?

In books and articles published over the last ten years, I have taken a range of approaches to this question, probing some of the sources of Twain's fiction, the processes by which he transmuted those sources into art, and the responses his work elicited. As editor of The Oxford Mark Twain, I invited some of the leading writers of our time to respond to Twain as one artist to another, continuing the cultural conversation, and I asked key Twain scholars to set each work in its biographical, social, and cultural context. Although I will refer to and build on what I learned in these earlier investigations, I take a rather different tack in this book.

The interconnected meditations that make up *Lighting Out for the Territory: Reflections on Mark Twain and American Culture* explore some of the challenges that confront us as we follow Twain into the territory that he made his own—and ours. In Chapter One, "The Matter of Hannibal," I invite the reader to accompany me on a visit to Hannibal, Missouri, where I explore how that community shaped Mark Twain's work and how it honors his memory today. Chapter Two, "Excavations," describes my efforts to recover and vivify some "missing" chapters of the past. Chapter Three, "Ripples and Reverberations," charts how Twain continues to shape the very texture of our lives in the United States and around the world, tapping into the zeitgeist of his time and ours in ways we are still uncovering. In the epilogue I consider some of Twain's legacies for the twenty-first century.

What does the "use" we make of Mark Twain tell us about ourselves? Twain is there for us, speaking to our condition, as the Quakers say, no matter what that condition may be at any point, no matter how drastically it may change over time. The diverse and frequently competing images conjured up by the name "Mark Twain" were often carefully crafted and exploited by Sam Clemens himself. Twain spent a lifetime first projecting and then undercutting one image after another: a funny man with a talent for literature of the "low" sort; a serious author who despaired of being forever tarred with the "humorist" label; a satirist so subtle his meanings were often missed; a polemicist so direct his messages were often pointedly ignored. There is something for everyone in Mark Twain's opus: moral outrage, scintillating silliness, materialism, antimaterialism, nostalgia, antinostalgia, conformity, iconoclasm, technophilia, technophobia, exuberance, and bleak despair. The Twain we claim as our own reveals much about who we think we are—and who we want to be. In virtually any part of the world where there is literacy and printing, his works have been translated, read, and taught, yielding up to readers of vastly disparate geographies, economies, and educational levels a menu of stunningly varied tastes and textures. "Aren't you ready to move on to something else?" a friend asks, certain I should be bored by now with Mark Twain. The question makes no sense to me.

1

THE MATTER OF HANNIBAL

June 20, 1995. The summer sun shot through the window in blinding flashes as my plane approached the runway for a landing. I shielded my eyes. As the Fokker 100 touched the ground, I recalled the conversation I'd had eight months earlier with Masako Notoji, an ebullient professor of American Studies from the University of Tokyo, who was studying American theme parks and historic sites. She had been regaling me with stories of the half-dozen historic sites in the United States she'd just visited, one of which was Hannibal, Missouri. I confessed I had never been there. She was incredulous. "You haven't been to Hannibal? And you work on Mark Twain?" "But you must go," she scolded. "It's that simple: you *must* go."

The "Matter of Hannibal," as Henry Nash Smith called the world of Twain's youth and the world of *Tom Sawyer* and *Huckleberry Finn*, would preoccupy Twain throughout his career as a writer. Sometimes the slavery that helped make that world what it was hovered at the periphery of Twain's awareness, while at other times it was at the center.

Repeatedly Twain returned to events and scenes rooted in his Hannibal past, held them up to the light, turned them to see them from new angles, allowing them to cast fresh shadows. People and scenes that impinged on his consciousness in the present influenced what he recalled from that past, how he viewed it, and how he shaped it (consciously and unconsciously) into art.

Vivified through Mark Twain's imagination, Hannibal would become the scene of archetypal innocent idylls of childhood, the quintessential hometown. But it would also become a flash point of guilt, an emblem of bad faith and corruption, of moral rot, of barbarism—the underside of an arcadia that was innocent *only* in imagination.

When the well of his inspiration ran dry, as it did periodically, Twain often found that a quick detour back through the scenes of his childhood allowed it to fill again. But he could rarely predict the train of associations these returns would set in motion. Fragments of the past that had been muted or forgotten or buried would unexpectedly jump out at him. In Bombay in the 1890s, for example, the sight of a German abusing a servant vividly called up the chilling image of a slave in Hannibal being murdered by his master for some trifling offense. Daydream and nightmare would jostle one another, vying for primacy. The only constant linking all of these visions was Hannibal itself—real only as his mind chose to recall it, yet there, in some sense, in actuality, to return to in body as well as in spirit.

Twain did return physically—seven times, in fact—the last time being in 1902, when he was celebrated as a conquering hero. As a writer, however, he returned to Hannibal many more times. Hannibal would be the St. Petersburg of *The Adventures of Tom Sawyer* and the sequels, and of *Adventures of Huckleberry Finn* (where parts of it may have made their way into Bricksville and Pikesville as well). It would appear in the shape of Dawson's Landing in *Pudd'nhead Wilson* and Eseldorf (literally "Assville") in *The Mysterious Stranger*. It would peek out in various guises from the pages of *Life on the Mississippi*, *A Connecticut Yankee in King Arthur's Court*, "The Private History of a Campaign That Failed," and *Following the Equator*, as well as the autobiographical dictations and the posthumously published "Tom Sawyer's Conspiracy" and "A Scrap of Curious History." Dimensions of Hannibal's complacency, pretentiousness, and bad faith would surface in "The Man That Corrupted Hadleyburg" and "My First Lie and How I Got Out of It"—a suggestive but incomplete list.

Mark Twain's Hannibal is a palimpsest that yields diverse and often contradictory meanings. It is also a microcosm of America itself—its promise and its potential, its guilt and its shame. In *The Burden of Southern History*, C. Vann Woodward observed that "the tragic aspects and the ironic implications" of American history have been obscured by "the national legend of success and victory and the perpetuation of infant illusions of innocence and virtue." Hannibal eventually came to evoke, as no other single locale in the nation would or could evoke, both the innocence and the irony of American history.

* * *

As I pulled my rental car out of the St. Louis airport, heading northwest on Interstate 70, I recalled the view of Twain's snaking river from the plane. Meachum's river, too, I thought. Hannibal was my destination, but for the moment I was still in John Berry Meachum's town and I let my thoughts wander to this remarkable figure as St. Louis receded behind me. Born a slave in Virginia in 1789, Meachum became a skilled carpenter, cabinetmaker and barrel maker, purchasing his own freedom and that of his father before marrying and starting a family with a woman who was still a slave. When her master moved to Missouri, Meachum followed, soon purchasing his wife's and children's freedom as well and settling them in St. Louis. After establishing himself as a successful steamboat entrepreneur, Meachum bought twenty slaves, taught them a trade, employed them in his barrel-making business, and allowed them to purchase their freedom from him with the wages he paid them. He also became an important religious leader: ordained as a Baptist minister in 1825, he founded (in collaboration with New England evangelist John Mason Peck) the first black Protestant congregation west of the Mississippi. But it was his commitment to education that indelibly etched him in my memory.

In the late 1820s, in violation of a city ordinance, Meachum founded a clandestine school in the basement of his church on Third and Almond Streets, the first school for blacks in St. Louis. Under the cover of receiving religious instruction, his pupils, both slave and free, were taught to read and write and were encouraged to view education as the key to their future success. In 1847, however, unnerved by escalating fears of slave uprisings, the Missouri state legislature did away with Meachum's school (or so they thought), passing a law stipulating that no one could "keep or teach any school for the instruction of negroes or mulattoes in reading or writing in this STATE." The punishment for those who broke the law and couldn't pay the fine was public whipping. John Berry Meachum was undeterred. In a move that deserves to be remembered in the history of American education as a masterpiece of ingenuity, Meachum outfitted a steamboat with books and anchored it in the middle of the Mississippi River, where it was subject to federal but not state laws. Students were ferried to the boat by skiff, taught reading, writing, and arithmetic all day, and ferried back to shore in the evening. Meachum's floating "Freedom School" continued until his death in the late 1850s.

As I turned off the interstate onto famous Highway 61, I thought about Meachum's creativity and courage—and about his obscurity. I had come across him by accident in my research. No monument or statue is erected to his memory, and he never makes it into American history books, Missouri histories, or histories of education. Why?

I recalled my graduate seminar on the African-American press last semester at the University of Texas at Austin. One of my students, Bruce Wilson, a black veteran whose knowledge of Texas black history is longstanding and rich, had chosen to investigate an incident that had received widespread attention in the local paper and the national black press when it first happened but somehow had not made it into standard histories of the state. John. R. Shillady, the white secretary of the NAACP, had been brutally beaten in broad daylight during his visit to Austin in 1919 in an attempt to discourage his efforts on behalf of the NAACP in Texas. One of Shillady's attackers, a well-known county judge, bragged, "I told him our Negroes would cause no trouble if left alone. Then I whipped him and ordered him to leave because I thought it was for the best interests of Austin and the state." When the NAACP national office asked the governor of Texas what was being done to punish the offenders, he replied by telegram, "Shillady was the only offender in connection with the matter." But it was the names of these powerful and visible leaders in Texas politics that sent shock waves around the classroom when Bruce Wilson spoke them: Pickle and Hobby. The county judge and the governor involved in this 1919 incident had the same last names as two highly respected contemporary elder statesmen of Texas who had firmly held the reins of power in the state until their retirement in the early 1990s, one as a U.S. congressman, the other as lieutenant governor. It was not, alas, some strange coincidence: they were, respectively, the grandfather of one and the father of the other.

I began to see why there was no monument to John Berry Meachum. One cannot honor the achievement of blacks who fought against their oppression without shining a glaring light on that oppression and on those who perpetrated it. One has to be willing to finger the first families of the state, longtime political dynasties, and good churchgoing folks like

Aunt Sally and Uncle Silas in *Huckleberry Finn* as complicitous in an obscene and barbaric system. Celebrating the bravery and heroism of a Meachum required acknowledging the baseness of the good citizens of Missouri who threw those legal obstacles in his path. Small wonder that for many whites it seems simpler—and infinitely safer—just to forget the whole thing.

* * *

I checked into the Best Western Hotel Clemens in the late afternoon. When I realized I was a block from the Mark Twain Historic District, I decided to go for a walk. Shortly after I turned left onto Hill Street, I spotted a historic marker that read:

> HERE STOOD THE BOARD
> FENCE WHICH TOM SAWYER
> PERSUADED HIS GANG TO
> PAY HIM FOR THE PRIVILEGE
> OF WHITEWASHING. TOM
> SAT BY AND SAW THAT IT
> WAS WELL DONE.

Wait a minute, I thought. Tom Sawyer was a fictional character and was thus incapable of doing anything to an actual board fence that may or may not have existed at one time on Hill Street. And I knew that if Tom's fence allegedly "stood" on this spot some time in the past, at least ten more "Tom Sawyer's Fences" would miraculously materialize when Tom Sawyer Days took place some two weeks from today. I responded to the historical markers in Twain's hometown with the same healthy skepticism Twain himself had trained on the "facts" related by tour guides in *The Innocents Abroad*:

> We find a piece of the true cross in every old church we go into, and some of the nails that held it together. . . . I think we have seen as much as a keg of these nails. Then there is the crown of thorns; they have part of one in Sainte Chapelle, in Paris, and part of one, also, in Notre Dame. And as for bones of St. Denis, I feel certain we have seen enough of them to duplicate him if necessary.

I smiled to myself when I realized the kind of tourist I'd become: the kind Mark Twain had trained me to be.

I had only a few minutes in the Mark Twain Museum Annex before it closed, so I watched the biographical slide show and then hurriedly flipped through the rack of glass-covered boards displaying clippings and memorabilia associated with Twain and his world. Two items caught my interest. One was a 1935 poem by Edgar Guest composed for the Twain centennial.

> Down in Hannibal, Missouri, they're living once again
> All the countless happy memories of a boy they called Mark
> Twain. . . .
> . . . .
> Down in Hannibal, Missouri, young and old with eyes aglow
> Are remembering a baby born one hundred years ago.
> They are pointing out the places where that little fellow played,
> And the haunts he made immortal by some boyish escapade. . . .

The other was a page from the *St. Louis Republican* of 1849, displayed with a legend stating that this issue of the newspaper "reached Hannibal when Samuel Clemens was 14 years old." Among the ads for Perry Davis's painkiller medicine, the circus, and various theatrical productions were these:

NEGROES for SALE A strong, healthy woman 35–40 years of age, a first-rate cook, ironer and washer. Speaks French and English, and her two children, a boy seven years, and a girl three years old. Also, a sprightly 11 year old girl, all from the country. A first-rate woman cook, washer and ironer, for sale, not to leave the city. Apply at 104 Locust St.

NEGRO GIRL FOR SALE A likely young negro girl, about 14 years old—sold for no fault. The owner having no use for her, would prefer to sell to a resident of the state. Apply on Broadway, first brick house south of Howard St.

The museum was closing. I'd have to save the rest of the exhibits for the next day. As I made my way out into the street, I wondered how Hannibal was going to reconcile the world of "countless happy memories" of carefree boyhood with the world embodied in those chilling ads.

* * *

Exiting onto Hill Street, I walked past the Mark Twain Boyhood Home, John Marshall Clemens' Law Office, the Haunted House, the Becky Thatcher House, and the Twainland Express Depot. I turned left on Third Street and found myself in what looked like the commercial

1. *The Mark Twain Family Restaurant and the Twainland Express depot at the corner of Third and Hill Streets in Hannibal, Missouri (Photo courtesy of R. Kent Rasmussen)*

district of any town of comparable size: a real estate office, a business supply store, a photography studio, a car stereo and cellular phone store, an electric heating equipment company, a radio station, a luncheonette, and a newspaper office. This had been a commercial area in Twain's day as well, I recalled. I stopped when I reached the corner of Third and Center Streets. When Sam Clemens was twelve, a businessman named John Armstrong traded hay, grain, and slaves at Melpontian Hall on this corner.

Slavery in Hannibal may not have been "the brutal plantation article," but it was slavery nonetheless, with the all too familiar mix of pain and powerlessness. Emma Knight of Hannibal was born a slave near Florida, Missouri, Mark Twain's birthplace. When she and her sisters outgrew the shoes their master gave them only once a year, they had to go barefoot. "Our feet would crack open from de cold and bleed. We would sit down and bawl and cry because it hurt so," she told an interviewer years after freedom had come. Her family had been separated, her father sold at auction—and not simply to settle an estate. "My father was took away. My mother said he was put upon a block and sold 'cause de master wanted money to buy something for de house." Clay Smith, another slave

2. *Sign directing tourists to the "Haunted House on Hill Street" Wax Museum in Hannibal, Missouri (Photo courtesy of R. Kent Rasmussen)*

from Hannibal, recalled that her aunt Harriet "was sold on de block down on Fourth Street right here in Hannibal."

The slave trading at Melpontian Hall—about four blocks from the house where the Clemens family lived—was so repugnant to the Moores, a newly arrived family in town, that they packed up and moved back to Wisconsin after a very brief stay. Yet the disdain in which the citizens of Hannibal allegedly held the slave trader was not so strong as to dissuade them from using Melpontian Hall as their voting place on election day.

All slaves were vulnerable to being sold away from friends and family. Indeed, as Twain tells us, his own father was responsible for one such sale, having exiled a slave named Charley "from his home, his mother, and his friends, and all things and creatures that make life dear." In 1842 John Marshall Clemens, who had received the slave in settlement of a long-standing debt, took Charley with him on a trip to collect $470 he was owed by a man in Mississippi. John Clemens found the financial trials of the Mississippi man so moving that he "could not have the conscience" to collect the debt (as he wrote home). But he had no qualms about selling Charley down the river for about forty dollars' worth of tar—the same amount that the king and the duke got for Jim when they sold him in the novel Twain would write some forty years later.

That the Deep South held no monopoly on cruelty as far as slaves were concerned is clear from Twain's own recollections. At age ten, in 1845, on one of Hannibal's main streets, he had watched a white master strike and kill a slave with a piece of iron, a memory that came back to him in Bombay. "I knew the man had a right to kill his slave if he wanted to, and yet it seemed a pitiful thing and somehow wrong, though why wrong I was not deep enough to explain. . . . Nobody in the village approved of that murder, but of course no one said much about it." On another occasion he recalled the community's response to the death of a slave at the hands of a white overseer: "Everybody seemed indifferent about it as regarded the slave—though considerable sympathy was felt for the slave's owner, who had been bereft of valuable property by a worthless person who was not able to pay for it." (The jarring intrusion of the fact that the murder of the slave left the owner "bereft of valuable property" resonates with the dry denouement of *Pudd'nhead Wilson*: "Everybody granted that if 'Tom' were white and free it would be unquestionably right to punish him—it would be no loss to anybody; but to shut up a valuable slave for life—that was quite another matter. As soon as the Governor understood the case, he pardoned Tom at once, and the creditors sold him down the river.")

Although the white citizens of Hannibal may have persuaded themselves that their "mild domestic slavery" was more humane than "the brutal plantation article," the efforts of slaves to escape at great personal risk and the fears of their owners that they would succeed belie the view that "as a rule our slaves were convinced and content." Emma Knight recalled her mistress's efforts to intimidate the slaves: "Mistress always told us dat if we run away somebody would catch and kill us. We was always scared when somebody strange come." Clay Smith recalled that "Father run away to Illinois during the war and we ain't never saw him again." Twain recalled from his early childhood in Florida, Missouri, hearing the "loud and frequent groans" of a runaway slave brought into the town "by six men who took him to an empty cabin, where they threw him on the floor and bound him with ropes." In 1847, when Twain was eleven, a runaway slave who belonged to a man named Neriam Todd swam across the river and hid in the swampy thickets of Sny Island, on the Illinois side of the Mississippi. A boy of Twain's acquaintance, Benson Blankenship, found him and brought him scraps of food instead of giving him up for a reward. (His behavior would become a model for aspects of Huck's behavior in *Huckleberry Finn*.) Some woodchoppers chased the slave into a part of the swamp called Bird Slough, where he disappeared; several days later, Sam Clemens and a few of his friends who had crossed the river to fish and hunt for berries found the slave's mutilated body. Yet courage persisted in the face of cruelty and danger. Indeed, slaves escaped with just enough frequency that insurance companies advertised policies to help protect slave owners from the financial loss involved.

* * *

There was something else the slaves had that pain and powerlessness and poverty didn't manage to extinguish: a rich and creative oral tradition. A young Sam Clemens who as yet knew nothing of his future calling listened to it every chance he got. The slaves didn't tell this attentive little white boy how much they suffered: stories like those Emma Knight and Clay Smith shared with interviewers years after slavery ended were not for his ears. What they did let him hear were ghost stories and satirical orations so masterfully constructed and delivered that he would remember them all his life. He was tremendously struck by the storytelling talents of Uncle Dan'l, a slave at his uncle's farm in Florida, Missouri, whose tales he was privileged to listen to every night in the summer. In a letter Twain wrote about him in 1881, he recalled the "impressive pauses and eloquent silences" of Uncle Dan'l's "impressive delivery." Twain would also recall the rhetorical performances of Jerry, "a gay and impudent and satirical and delightful young black man—a slave, who daily preached sermons from the top of his master's woodpile, with me for sole audience. . . . To me he was a wonder. I believed he was the greatest orator in the United States." All his life Twain would emulate the lessons in storytelling and satire he learned from Uncle Dan'l and Jerry during his Hannibal childhood. These master talents, however, despite their consummate skill as artists, were still slaves, and as such they were just as vulnerable as Charley was to being sold and separated from the people they loved. Later in life—much later—Twain would comprehend what that meant.

As an adult Twain would remember the sight of "a dozen black men and women chained to each other . . . and lying in a group on the pavement, awaiting shipment to the southern slave market. Those were the saddest faces I ever saw." In *A Connecticut Yankee* he would describe another chained group of slaves: "Even the children were smileless; there was not a face among all these half a hundred people but was cast down, and bore that set expression of hopelessness which is bred of long and hard trials and old acquaintance with despair."

During his Hannibal boyhood, Sam Clemens did not challenge the social and legal norms that produced that despair. "In those old slave-holding days," he recalled,

> the whole community was agreed as to one thing—the awful sacredness
> of slave property. . . . To help steal a horse or a cow was a low crime,
> but to help a hunted slave . . . or hesitate to promptly betray him to a
> slave-catcher when opportunity offered was a much baser crime, and
> carried with it a stain, a moral smirch which nothing could wipe away.
> . . . It seemed natural enough to me then.

But the boy who found this state of affairs "natural enough" would ultimately come to hold a very different attitude.

> What is a "real" civilization? Nobody can answer that conundrum.
> They have all tried. Then suppose we try to get at what it is not; and
> then subtract the what it is not from the general sum, and call the
> remainder "real" civilization. . . . Let us say, then, in broad terms, that
> any system which has in it any one of these things, to wit, human
> slavery, despotic government, inequality, numerous and brutal punish-
> ments for crimes, superstition almost universal, ignorance almost uni-
> versal, and dirt and poverty almost universal—is not a real civilization,
> and any system which has none of them is.
> If you grant these terms, one may then consider this conundrum:
> How old is real civilization? The answer is easy and unassailable. A
> century ago it had not appeared anywhere in the world during a single
> instant since the world was made. If you grant these terms—and I
> don't see why it shouldn't be fair, since civilization must surely mean
> the humanizing of a people, not a class—there is today but one real
> civilization in the world, and it is not yet thirty years old. We made
> the trip and hoisted its flag when we disposed of our slavery.

This child of slaveholders, this member of a ragtag band of Confederate irregulars, would write what is arguably the greatest antiracist novel by an American: a book about a young boy who is oblivious to anything amiss in the moral universe of the grown-ups around him (as Clemens himself was) but who, despite his best intentions to the contrary, allows his "sound heart" to defeat his "deformed conscience," forging, in the process, one of the most memorable interracial friendships in literature.

What an incredible story Clemens' own story was: a young boy who accepts slavery as natural and right grows up to become a man who asserts that civilization began when slavery was abolished. Along the way he becomes the most famous American writer of his time—perhaps of all time. How, I wondered, would Hannibal dramatize this compelling saga?

As I made my way back along a dark North Third Street to the Best Western Hotel Clemens, I stopped at the corner of Hill to look once more at the river. A magical mile-wide ribbon of undulating liquid silver. A liminal space, once a fluid border between slave territory and free, powerful, mysterious, as fearful and unpredictable as nature itself. Twain's river. Meachum's river, too. Suddenly a twinge of doubt crossed my mind, but I banished it: Missouri could forget Meachum, but Hannibal could never forget Mark Twain. The Mark Twain Historic District stretched out in front of me, proof of the folly of the thought that had disturbed me for a moment. But then it gnawed at me again: Which Twain? Might they somehow have managed to forget the one that mattered most?

*   *   *

Back in my hotel room, I curled up with the small stack of books and articles I'd brought from home or acquired that afternoon and read up on local history. From Hannibal-born journalist Ron Powers' *White Town Drowsing* I learned that

> no one knows for sure who named the site Hannibal or why. The official histories of that region—that is, the white histories—are content to point out that the settlers of the early nineteenth century had a taste for classical allusions; and it is true that many of the smaller towns near Hannibal suggest this taste. . . . At any rate, the official histories seldom go further than identifying Hannibal as the famous Carthaginian general. One needs to check a few independent sources to confirm that Hannibal was a *black* Carthaginian general.

This fact, Powers observes, "has tended to legitimize a cherished but carefully guarded oral legend passed through the generations of the town's small community of black citizens." According to this legend, the earliest settlement of non–Native Americans in the area was

> a campsite struck sometime after 1804 by members of the Lewis and Clark expedition of the Louisiana Territory. Among the members of this expedition was an African slave, a former explorer and ocean pilot who had been taken into slavery by Thomas Jefferson himself, then sent west with Captain William Clark.

The slave, who resisted Clark's efforts to name him "York," insisted on calling himself "Hannibal."

> York—or Hannibal—was assigned to guard the expedition's campsite. Members of the party who ventured inland—so the legend goes—could always find their way back to the base by fixing upon the sight, or sound, of the fierce African who waved a flag from side to side and announced into the western continent, "This is Hannibal! This is Hannibal!"

Powers finds the legend "appealing—the town, at its earliest nascence, bawling out its being in a human voice."

Some two decades after Sam Clemens left Hannibal, the sleepy river town of his childhood was transformed into a bustling commercial hub, the fourth-largest lumber center in the nation. Starting in the 1870s, lumber from Minnesota and Wisconsin was floated down the river by raft to Hannibal, where it was processed in local sawmills and then shipped by freight train to Kansas, Nebraska, Colorado, and points west. Great fortunes were made and Hannibal's lumber barons built imposing mansions.

But in 1901 Hannibal's lumber-hauling season ended early because the river had dropped too low. In addition, the white-pine forests of Minnesota and Wisconsin had been heavily depleted, and soon newly built railroads and sawmills allowed the lumber that was left to be processed locally in the northern states and shipped directly, instead of being sent downriver to Hannibal. By 1903 the lumber industry had largely abandoned Hannibal. Shoe factories, a cement plant, cigar factories, and other manufacturing enterprises remained, but in time many of these would go the way of the lumber business.

There was one resource Hannibal had that would prove to be infinitely renewable: its status, as Hannibal historians Hurley and Roberta Hagood put it, as "the town which furnished the background for the escapades of Tom Sawyer and Huck Finn." Hannibal is not "on the way to" anywhere. It is a drive from any major airport, and passenger trains don't run there. " 'Mark Twain's Boyhood Home,' 'America's Home Town,' and the 'Best Known Little Town in the World' " nonetheless manages to attract as many as 350,000 visitors a year. In 1992 tourists pumped over thirteen million dollars into Hannibal's economy.

"Missouri claims Mark Twain for its very own. . . . The commonwealth writes his name upon its role of sons distinguished and watches with maternal pride his globe-girdling career," wrote Walter Williams in *Five Famous Missourians* in 1900. But at the time of Twain's death in 1910, the little white house he had lived in on Hill Street had fallen into dis-repair, and there were plans to turn it into a butcher shop. In 1911 Mr. and Mrs. George Mahan of Hannibal intervened, purchasing the house and presenting it to the city. In the years that followed, they gave the city the statue of Huck and Tom that still stands at the foot of Cardiff Hill (the first statue in the country of purely fictional characters), and they placed historical markers at various sites. In short, they laid the groundwork for a tourist industry that would eventually be responsible for more than fifteen hundred local jobs.

The Mark Twain Museum (originally housed in a bank) was dedicated in 1935 by Twain's daughter, Clara Clemens Gabrilowitsch. The 1935 centennial of Twain's birth was celebrated with concerts, parades, and a massive biographical pageant presented on three nights in June on the new high school football field. The pageant—"Mark Twain's First One Hundred Years"—featured a shooting comet, a horse-drawn carriage, and a cast of eleven hundred local citizens. President Roosevelt dedicated the Mark Twain Lighthouse, illuminating it with the flip of a switch in the White House, an event that was broadcast live on radio across the nation. The Mark Twain Bridge, linking Missouri to Illinois, was completed and dedicated the following year. Over the next fifty years, 6.5 million tourists would visit the town.

At the sesquicentennial celebrations in 1985, instead of a pageant about Mark Twain's life there were Twain impersonators, an evening of readings by an actress impersonating his mother, appearances by "Tom" and "Becky," and the daily downtown antics of various citizens dressed up as "Hannibal the Frog." President Reagan proclaimed November 30 (Twain's birthday) Mark Twain Day and the postal service issued a spe-cial Mark Twain/Halley's Comet U.S. Aerogramme in December. In the slogan selected for the sesquicentennial—"Hannibal's Jumping"—there was a nod both to the frog that gave Twain his first national visibility and to the aspirations of the town that wished to capitalize on his fame. But a combination of inexperience, mismanagement, and political imbro-glio prevented the "Sesqui" from making the huge profits its enthusiasts had predicted and left the city with some embarrassing debts instead.

I closed the books, opened the curtains, and looked out the window. I recalled the story about the great Argentine writer Jorge Luis Borges that I had just read in the chapter on Hannibal in William Zinsser's *American Places*. When he was eighty-three, Borges agreed to lecture at Washington University in St. Louis on the condition that his hosts take him to Hannibal. Twain's work—particularly *Huckleberry Finn*—had cap-

3. Pictured in front of the Mark Twain Boyhood Home in Hannibal, Missouri, during the Mark Twain Centennial in 1935 (from left to right), George A. Mahan, President of the Missouri State Historical Society and narrator of the Centennial pageant; Morris Anderson, Chairman of the Hannibal Mark Twain Centennial Committee; Clara Clemens Gabrilowitsch, Mark Twain's daughter; Hannibal youths dressed as "Huck Finn" and "Becky Thatcher;" sculptor Walter Russell of New York; Hannibal school boy dressed as "Tom Sawyer," and Nina Gabrilowitsch, Clemens' 18-year-old granddaughter (Photo courtesy of the Mark Twain House, Hartford, Connecticut)

tured his imagination as a child and sustained him as an adult. Frail and nearly blind, he insisted on making the two-hour trip to Twain's hometown. When he got there it became clear that there was really only one thing he wanted to do: put his hand in the Mississippi River. He reached down and did just that. The river, he said, was the essence of Twain's writing. He had to touch it.

I watched the moon throw handfuls of dancing stars—whole constellations of them—at the shimmering water. I was not convinced the river was the source of Twain's art. But like Borges I had felt the need to visit Hannibal, sure that coming here would somehow take me closer to the wellspring of Twain's imagination. Would it? I wondered, as I drifted off to sleep.

* * *

I woke up as the sun was coming out, dressed, and walked down Hill Street to the river. The silver ribbon of the night before had become an unremarkable gray-brown swath of water. But it intrigued me nonetheless: it was the same view Sam Clemens had looked out at each morning. Mark Twain's lapidary prose would make the dawn on that river in *Huckleberry Finn* a kind of ur-dawn that all other literary dawns would have to live up to:

> Not a sound, anywheres—perfectly still—just like the whole world was asleep, only sometimes the bull-frogs a-cluttering, maybe. The first thing to see, looking away over the water, was a kind of dull line— that was the woods on t'other side—you couldn't make nothing else out; then a pale place in the sky; then more paleness, spreading around; then the river softened up, away off, and warn't black any more, but gray. . . .

I took a deep breath, filling my lungs with the moist air of the Missouri summer. The rancid fishy smell nearly toppled me. Huck had warned me, but I had simply forgotten:

> the nice breeze springs up, and comes fanning you from over there, so cool and fresh and sweet to smell, on account of the woods and the flowers; but sometimes not that way, because they've left dead fish laying around, gars, and such, and they do get pretty rank.

I walked along the river for a while. Behind me lay "the white town drowsing in the sunshine of a summer's morning, . . . the streets empty or pretty nearly so." I took off my shoes and waded in the muddy coolness of "the great Mississippi, the majestic, the magnificent Mississippi, rolling its mile-wide tide along, shining in the sun." I gazed at "the dense forest away on the other side; the 'point' above the town, and the 'point' below, bounding the river-glimpse and turning it into a sort of sea, and withal a very still and brilliant and lonely one." At this juncture, Twain tells us in *Life on the Mississippi*, the stillness of the drowsing town is broken by the joyful shout of a black man, "a negro drayman, famous for his quick eye and prodigious voice," who spots "a film of dark smoke . . . above one of those remote 'points' " and instantly "lifts up the cry, 'S-t-e-a-m-boat a-comin'!' and the scene changes! . . . and all in a twinkling the dead town is alive and moving."

The famous "negro drayman" was a free black from Virginia named

John Hannicks, who lived in Hannibal with his wife, Ellen, and their three children. In addition to the distinction of usually being the first to spot a steamboat's approach, Hannicks was known for his helpfulness. In 1851 the *Hannibal Courier* praised the "exertions of good-humored 'JOHN,' the Drayman, in turning out with his dray and hauling water" to the scene of a fire. Mark Twain remembered him not only for his booming voice but also for his storytelling and his ready laughter. In a notebook entry dated 1887 he referred to "John Hanicks' [sic] laugh." In another notebook entry ten years later, he alluded to the drayman's "giving his 'experience,' " but the memorable story he must have told has not been preserved. Twain commented on Hannicks again in the jottings known as "Villagers of 1840–3," composed in 1897, forty-four years after Twain left Hannibal.

> John Hannicks, with the laugh. See black smoke rising beyond the point—
> "Steeammm*boat* a coming!" Laugh. Rattle his dray.

I tried to imagine John Hannicks's laugh that morning on the river. I tried to imagine his "prodigious voice." During the slide show at the museum annex the day before, when that passage from *Life on the Mississippi* came up in the narration, John Hannicks's famous "Steeammm-boat a-comin' " was there, but it was not attributed to anyone in particular, and the reference to the "negro drayman" had been left out.

Hannicks was a free man of color, and like other free blacks in Missouri, he had to carry a license to live in the state at all times and was obligated to post bond if he wanted to travel to another county. In the "South in my own time," Hank Morgan would recall in *A Connecticut Yankee,* "hundreds of free men who could not prove that they were free-men had been sold into life-long slavery." Hannicks ran a huge risk any time he left home without his license. He was required to register his children at the county court when they turned seven, and to have them bound out as apprentices and servants at that age, permitting the state "to enslave them by another name." What had it been like, I wondered, to be free and black in a slaveholding town like Hannibal? In addition to Hannicks and his family, more than thirty other free blacks lived in Hannibal during Twain's childhood. How was their history remembered and preserved?

As the morning sun played on the water, the empty streets began to show signs of life. Merchants unlocked their doors. A tour bus unloaded its passengers. I made my way back up Hill Street, arriving at the Mark

Twain Boyhood Home and Museum when it opened at eight. I peered into the bedroom Sam had shared with his brother Henry, the family's kitchen, the parlor—all beautifully restored in 1990 when each room was fitted with reconditioned nineteenth-century hardware, painted with the kind of paint that was used in Hannibal when Clemens lived there, and furnished with authentic period pieces resembling those that the Clemens family was known to have owned. Although the rooms were sealed behind glass to protect them from temperature change and moisture, I experienced the full blast of the day's building heat as I walked through the house on open-air metal platforms that helped decrease wear and tear on the structure itself. Tape recordings that played over loudspeakers positioned in front of each room provided basic information about how that room was used and connected it to specific scenes in *Tom Sawyer*. The museum I entered after walking through the home featured, among other things, a glass case containing marbles that had been found during excavations of the site—marbles Sam Clemens himself may have played with.

The restoration was impressive, I told curator Henry Sweets III when he met me in the garden outside for our interview. He beamed. It was just the beginning, he said. All kinds of new plans were in the works for the Mark Twain Boyhood Home and Museum—particularly now that Hannibal was no longer vulnerable to the flooding that had periodically devastated the downtown area for more than a century. A measure of how seriously Hannibal takes historic preservation is the five-hundred-year flood wall built to protect the historic district in 1992. This is a town that wants the past to last. "The flood protection . . . kept us high and dry in the flood of 1993. Things are looking good for Hannibal in the future," Sweets said. Building designs for the new museum had recently been approved, and preliminary fund-raising efforts had succeeded to the tune of well over a quarter of a million dollars. Animated and excited, he described a museum whose first floor would feature a series of rooms or alcoves, each devoted to one of Twain's major books. At the rear a grand staircase similar to those found on steamboats would take the visitor up to a pilothouse high enough to afford a view of the river. Rotating exhibits would focus on Hannibal's railroading heritage, cigar-making heritage, and shoemaking heritage. New buildings might be added as well, such as a nineteenth-century print shop and a schoolroom. The long-term plans involved encouraging more craftsmen to relocate to the downtown area adjacent to the historic district. Some had begun to do so; a glassblower and a potter were there already. He said it would be nice to have a metalworking shop. "We want the whole downtown rejuvenated," transformed into "a place visitors will want to come to for trips." Close to eight million people had visited the Mark Twain Boyhood Home and Museum, 120,000 in 1994 alone, Sweets said. He hoped the expansion would lure even more.

"Is Twain's antiracism known here? Is it taught here?"

"It isn't brushed under the rug," Sweets said. "It just isn't approached. . . . The tie to Mark Twain for Hannibal is *Tom Sawyer*. The connection people feel is really through *Tom Sawyer* rather than through any other of the writings."

Tom Sawyer Days was testimony to the power of that book over this town. Two weeks hence, Sweets noted, Hannibal would celebrate its fortieth Tom Sawyer Days. Each year since 1956 a seventh-grade boy had been crowned "Tom Sawyer" during the festivities, and a seventh-grade girl had been crowned "Becky Thatcher." Although there is only one "official" Tom and one "official" Becky each year, the four runners-up in each competition work pretty hard as well. "We have days when we need multiple Toms and Beckys," Sweets explained. "Suppose there are two parades in two communities and both want a Tom and a Becky. Suppose you have two businesses opening the same day. Also, you don't want to use the same ones all the time because there are so many things they use the Toms and Beckys *for*. Suppose there's a business opening on a Thursday, and you need to pull a Tom or Becky out of school. Well, you don't want to pull the same one all the time or you'll have problems." This year all forty of the official Toms and official Beckys since 1956 had been invited back to Hannibal for a special dinner on July 3. Were any of them in town yet? I asked. Sweets rattled off the names of eight who had never left Hannibal to begin with. Becky Thatcher of 1964, in fact, was head of the Visitors and Convention Bureau, which was next to my hotel. He was sure she'd be glad to talk to me. I wrote down her name in my notebook.

   *   *   *

After my conversation with Henry Sweets, I walked across the street to the Mark Twain Book and Gift Shop, an establishment that had no connection other than proximity to the Mark Twain Boyhood Home and Museum. What would Twain have thought of *Tom Sawyer, Huckleberry Finn, The Prince and the Pauper* and *A Connecticut Yankee* nestled up against the complete works of William Shakespeare—a writer Twain alternately competed with and made fun of—in a bookstore in Twain's hometown? Since he was in the bookselling business himself, not just as a writer but as a publisher, Twain would have understood the owner's impulse to carry, alongside Twain's greatest novels, a book entitled *365 Pies You Can Bake*. After all, in 1893 one of the books Twain brought out as he struggled unsuccessfully to save his publishing company from bankruptcy was Alexander Filippini's *One Hundred Desserts*. I purchased a handful of postcards and several copies of a facsimile reprint of *Harper's Weekly* from April 30, 1910—nine days after Twain's death—before breaking down and buying Ernest Matthew Mickler's *White Trash Cooking*. The amiable proprietor, Martha Adrian, rang up the sale and showed me around when I told her that I was writing about Mark Twain and Hannibal, and wanted to know which souvenirs were the most popular.

We took a leisurely stroll among the densely cluttered shelves as Adrian expanded on her wares. Items for sale included a bust of Mark Twain, a framed Tom Sawyer stamp showing a famous Norman Rockwell illustration, a music-box reproduction of the Mark Twain Lighthouse on Cardiff Hill, prints of steamboats, original pages from turn-of-the-century magazines with material pertaining to Twain, cast-iron "Aunt Jemima" and "Uncle Moss" banks, and plaster figurines made in China depicting black children eating watermelon, tying their shoes, or sitting on the toilet. There were rows of collectible spoons and thimbles featuring images of Mark Twain and Tom and Huck and Becky, and magnets with Mark Twain's picture on them. "In the book section, I've tried to have all of his books," Adrian said. "The best-seller would be *Tom Sawyer*, after that *Huck Finn*." She gestured to the stacks of Confederate army caps on the shelf beneath the books. "We sell quite a few of those, but just as many Northern hats. Now on the bandannas, we sell more of the Confederate bandannas. It's a pretty popular item with kids now—they like to wear it on their heads." I asked about the "Confederate Generals" and "Union Generals" decks of cards. "Probably 'Confederate Generals' sell better," Adrian answered.

"What about the bullwhips?" I pointed to a rack of whips hanging next to us.

"That's just a popular item with little boys. They just for some reason like to snap a bullwhip. The [previous] owner had 'em when we bought the store. They sold real well. It's just a perennial thing."

"What are people looking for when they come here?" I asked. "What do they want to find in Hannibal? Why do they seek out Twain's boyhood home?"

It was "the small-town values that he had," Adrian said. "Most people, when they read things of his, they were either raised in a small town or they're interested in that type of thing. . . . A lot of people come here and say what a nice place this would be to live." Her customers include families on vacation, honeymooners, tourists from all over the United States and around the world and, of course, hundreds of groups of schoolchildren.

"What's the most popular souvenir when school groups come?"

"Bullwhips."

* * *

"What did it take to be Becky?" I asked Faye Bleigh, winner of the "Becky Thatcher" competition of 1964.

"To be as nice and as pleasant and friendly as I possibly could because I was an ambassador for Hannibal," she said.

Bleigh, director of the Hannibal Visitors and Convention Bureau, is still nice and pleasant and friendly and she is still an ambassador for Hannibal. With obvious pride she took out a copy of an April 1995 *Life* magazine and handed it to me. *Life* had called Hannibal "the best place to be from sea to shining sea" over the Fourth of July, when the famous Tom Sawyer Days would take place and the 1995 Tom and Becky would be crowned. "The festival is an orgy of wholesomeness," *Life* declared. The full-page color picture of last year's fence-painting contest burst with the energy and exuberance of the barefoot adolescent contestants.

The seventh-grade girls vying to be Becky needed a lot more than a friendly smile these days: yards of eyelet and ruffles; also gloves, a bonnet, an umbrella, pantaloons, and shoes. "The costumes are made by the mother or someone paid to make it," Bleigh said. "The costume today probably costs about three hundred dollars to make. . . . It's very elaborate. It has gotten to be such a big deal that when they announce the Beckys they tell who made their dress." Girls who don't find the image of Becky Thatcher to their taste—and Faye Bleigh's daughter was one—can compete in the Tomboy Sawyer contest, painting fences like the Tom competitors, participating in watermelon seed–spitting contests or bubble gum–blowing contests, and wearing patched jeans instead of ruffles and eyelet.

Becky and Tom are supposed to have read their Twain. Bleigh read *Tom Sawyer* in seventh grade, as all Hannibal youngsters do, but didn't read *Huckleberry Finn* until she got to college. During her years in Hannibal High, Huck's book wasn't taught there.

"Have you ever heard a Hannibal child question the wisdom of Huck's handing over the lead to Tom Sawyer at the end of *Huckleberry Finn*? Is anybody ever disturbed by the imprisonment of Jim?" I asked.

"Never," Bleigh replied. "I've never heard anybody upset over it." She returned to her favorite subject: Hannibal's mission. "It is our goal here not only to promote the boyhood years, but when visitors come to town we hope that we have developed Hannibal in such a way as to let them step back in time and experience the excitement and the magic of Twain's writings. That's what it's all about. That's our goal here. . . . We are *so proud* that a famous American author lived in our town and we get to share him with generations and generations and generations. . . . We have a lot of fun just going back in time ourselves . . . especially in the historic district, where you have several blocks of mid-1800s buildings. You're going back in time. You get to see a piece of history. It's real."

I wondered where I would go to find a window on the African-American past in Hannibal. "Which are the sites that dramatize that?" I asked Faye Bleigh.

She looked flustered and said something about calling a gentleman connected with the NAACP. I rephrased my question.

"Which of the buildings in the historic district have evidence of there having been black Hannibal residents in the nineteenth century?"

"Well, I don't know what you mean by that," she replied.

I recalled the moving artifacts of nineteenth-century black life that I had seen in the African-American museum in Natchez two years before, and the traveling exhibit of photographic images of nineteenth-century black life in Mississippi that I had seen in London, and I remembered Twain's comment in *Life on the Mississippi* that when he returned to Hannibal in 1882 a black family was living in his boyhood home. I tried again. "Which sites make specific mention of African-Americans living in Hannibal?"

"I don't know whether we have a building that has that." Bleigh seemed puzzled.

I told her that one of the things I valued most about Twain was his moral growth, his ability to look back on an environment he didn't question in his childhood and write a devastating critique of it as an adult.

"I don't know whether he saw that when he was a boy or not," Bleigh said in measured tones.

"He didn't."

Her nervousness melted into relief. "See, and that's the only part we promote. His boyhood years. We don't [promote] the part where Huck and Jim are down the Mississippi. We promote only the little boy, when he played marbles, when he whitewashed the fence. That's the only part we promote."

When I asked about how Missouri history and the history of American race relations were taught in the local schools. Bleigh smiled broadly. She knew just the person I needed to see. He used to teach history. She was sure he'd be able to tell me everything I wanted to know. She phoned him on the spot: he'd be in his office for the rest of the afternoon, she said, and he'd be happy to see me. The former history teacher was Hannibal's mayor, Richard Schwartz.

*   *   *

The textbooks in the high school history classes Richard Schwartz taught in Hannibal in the 1970s and 1980s before he was elected mayor didn't cover the breakdown of Reconstruction as fully as he felt was necessary, so he had his students read supplementary books during the first five or ten minutes of class while he took attendance and did the mandatory paperwork. The supplementary material included Ida B. Wells' book documenting the epidemic of lynchings that ravaged the post-Reconstruction South and "covered the Jim Crow legislation, the *Plessy v Ferguson* case in a lot more detail than the textbook," he said as we chatted in his office in City Hall. "I thought it was important. In 1909 three blacks were lynched in Springfield, Missouri, and a medallion was printed to commemorate the event. It was grotesque." He wanted his students to know about that. Before the Civil War "there was a lynching here in Hannibal—a woman, a black woman. She was accused, I believe, of trying to encourage runaways."

The textbooks, Mayor Schwartz believes, often leave out some of the most important stories. "There's lots on the Spanish-American War, he said, "but there's no mention in books I've seen of the Philippine insurrection—the Philippine freedom fighters, the leader at the time, Aguinaldo. Mark Twain was totally against the war, but you don't hear of Aguinaldo or the Philippine insurrection. It lasted seven or eight years, cost much more in lives, ten times more than the Spanish-American War in money. But it's erased because it's a black mark on American history. It became a war of extermination. We may have depopulated entire islands and exterminated as many as half a million people. It's not something we're especially proud of. . . . Had we learned those lessons, had we taught that history, *maybe* we would have said: Vietnam? Do we want to get into a war where we're fighting freedom fighters, fighting guerrillas? I would hope that we as history teachers would be enthusiastic enough to teach the history—good and bad—all of it," he added "so that we can maybe as a people not make the same mistakes again."

The conversation turned to Mark Twain. He was so "sly the way he presented his point," so artful, the mayor said, that it was possible to read *Huckleberry Finn* and miss that point completely. It's "basically an antislavery, antiracism book—but if you read it the way you read *Tom Sawyer*, you never pick up on that."

I asked why there didn't seem to be any evidence anywhere in Hannibal of a celebration of Twain as an antiracist writer, the aspect of him that speaks most to our needs as a nation still struggling with racism.

The mayor paused a moment in thought. "That's a tough one," he said. "I'm not going to slough it off, but I'm going to tell you a story."

What followed were several stories—more than an hour and a half of them—vignettes from his own odyssey of racial awareness and chapters of Hannibal history with which they intersected. I liked this lanky, straightforward man in his forties, with his respect for history—his country's, his community's, and his own—who knew that history involved telling stories, and who had plenty to share. And I liked the way he took seriously questions that a politician of a different stripe might have thought impertinent or threatening.

When Mayor Schwartz was graduating from high school in 1969, his father and grandfather had a photographer show up to take a picture of three generations of Elks, assuming he'd be delighted that they had arranged for him to become a member. He refused to join: "The Elks in 1969 did not allow minorities, and I wasn't going to join a fraternal organization that didn't allow one of my best friends in high school—his name was Bob Frazier—to join." His father was angry, and his grandfather simply didn't understand. Schwartz had been friends with Frazier, who was black, since childhood. When he was around ten years old, Schwartz's grandmother called him up to her house on Hill Street one day and said, "I'm hearing these stories that you have a friend. . . . You need to be aware that people are talking." He ignored his grandmother's admonitions. Interracial friendships were frowned upon, Schwartz said, even in 1960s Hannibal.

Schwartz spoke of a range of incidents involving racism in Hannibal: the segregated past Hannibal had still not dealt with or eradicated (separate black and white American Legion posts, for example, still existed within blocks of each other downtown); the time one of his students became the local organizer of a Ku Klux Klan rally at Clemens Field

(Schwartz videotaped the rally—and the much larger International Committee Against Racism counterprotest—and brought the tape into class).

"How do you inoculate children against racism?" I asked.

He quoted the lyrics to a song from *South Pacific* about the need to be taught to love and to hate.

"Mark Twain seems to me a fantastic vehicle for that kind of education," I said. "I don't see him being used that way."

"You're right," the mayor said. "He's probably not being used that way. . . . And we're a perfect example of that, I suppose. By the way, did you hear what the Southern Baptist Convention did yesterday?"

I hadn't.

"They took a vote that slavery had been wrong. And they condemned racism. I heard it on the news. The Southern Baptist Convention. Amazing," he said.

"Well, better late than never."

I had asked whether he might direct me to some members of Hannibal's black community, and he helpfully gave me names and phone numbers—Hiawatha Crow, Reverend Ann Facen, Dixie Forte—and also suggested that I contact Hannibal historians Roberta and Hurley Hagood. His secretary had messages for him, and people were waiting to see him. Mayor Schwartz had given me much more of his time than I had had any right to expect, and I thanked him.

*   *   *

As soon as I left City Hall, I bought a copy of the *St. Louis Post-Dispatch* at a sidewalk vending box. There it was, right on the front page: BAPTISTS APOLOGIZE FOR SIN OF SLAVERY: BLACKS AND WHITES CLASP HANDS, VOW RECONCILIATION. The story by the *Post-Dispatch* religion writer, Patricia Rice, described the resolution approved the day before by the nation's largest Protestant denomination: "Southern Baptists apologized for the sins of their slave-owning founders and for more than a century of condoning racism, and white members clasped hands with black members to pray for forgiveness." The denomination's president, James B. Henry, and the highest-ranking black Southern Baptist, a pastor from Youngstown, Ohio, named Gary Frost, "embraced on the podium on the floor of the Georgia Dome" as the delegates, or messengers, who had just passed the resolution to work for racial reconciliation looked on. The meeting in Atlanta marked "the denomination's founding 150 years ago in Augusta, Ga. The denomination, with 15.6 million members, is the second largest in the country, after Catholics. Southerners founded it after they balked at New England Baptists' pronouncement that slave owners' 'hands were tainted with blood.'" Surfing through radio and television news broadcasts back in my hotel, I devoured any information I could find about the resolution and the vote taken on it the day before.

The head of the commission within the church that had drafted the resolution was the great-great-grandson of a slave owner. "I can't repent for him," he said. "I can't change his guilt before God for what he did. I can just apologize to those who are the descendants of the sufferers and for being part of a situation that we still live with today." Although the resolution was approved by a large majority of the mainly white crowd of twenty thousand in Atlanta, not everyone was happy about it. Convention leader Reverend Jere Allen of Washington, D.C., said he received a call from a Southern church deacon who angrily announced, "My ancestors had slaves and I'm very proud of it." And at the convention itself, a man from Baton Rouge objected from the floor that the resolution brought "discredit to those great men who founded this convention."

But the majority prevailed. On June 20, 1995, the denomination founded as a religious haven for slave owners had indeed approved a resolution that slavery "was an evil . . . from which we continue to reap a bitter harvest," and that the "racism which plagues our culture today is inextricably tied to the past." Acknowledging that "many of our Southern Baptist forebears . . . either participated in, supported, or acquiesced in the particularly inhumane nature of American slavery," and that "in later years Southern Baptists failed, in many cases, to support, and in some cases opposed, legitimate initiatives to secure the civil rights of African-Americans," the church that had come to be known as the denomination of the Ku Klux Klan resolved to "unwaveringly denounce racism in all its forms as a deplorable sin"; to "apologize to all African-Americans for condoning and/or perpetuating individual and systemic racism in our lifetime"; and to "genuinely repent of racism of which we have been guilty, whether consciously . . . or unconsciously." How do you do that? I wondered. Maybe the resolution had it right: acknowledging a past that was previously denied had to be the first step. For if we ignored the racism of the past, it was that much harder to recognize, let alone confront and transcend, the racism of the present.

I left my hotel room and headed for the river one more time before driving off to have a bite at Huck's Homestead Restaurant on Highway 61 and then catch the show at the Mark Twain Outdoor Theater at Clemens Landing. Walking down Hill Street, I heard a sharp report not precisely like any sound I had ever heard before—and then another. I found my path blocked by two teenage boys who stood in the street in front of the Mark Twain Boyhood Home and Museum cracking bull-whips. I talked with them for a few minutes. They were fourteen, and they were from Wisconsin. Finally I asked them what they thought the bullwhips would have been used for in Twain's day. Getting the horses to move faster, they said. I asked whether the whips might have been used on people, too. "I hadn't thought of that," one said. As I made my way toward the river, the boys swirled the whips on the ground in large, silent snakes and spirals.

*   *   *

The buffet dinner at Huck's Homestead Restaurant was vast, a feast that would have surpassed Huck's wildest fantasies: fried catfish, fried chicken, roast beef, green beans, smoked ham, pickled cucumbers, black-eyed peas, creamy corn pudding, mashed potatoes, gravy, hot rolls, bread pudding, and homemade pies. Outside the restaurant lively banjo music playing over loudspeakers drew the dinner crowd to an open-air amphi-theater built around a man-made lagoon that served as a stand-in for the Mississippi. The sets for "Reflections of Mark Twain" were impressive: an almost life-sized Hill Street and a fully detailed replica of a steamboat that chugged its way along the lagoon. "This delightful two-hour pageant is presented by twenty-five local actors who unfold the story of Mark Twain and his famous characters," read the brochure I had picked up in my hotel. "Episodes from *Tom Sawyer*, *Huckleberry Finn*, and *Life on the Mississippi* come alive in a very special way."

The first act was slow: a few scenes from *Life on the Mississippi* and a couple from *Tom Sawyer* (whitewashing the fence was played up big), weighed down by some insipid musical numbers. Things began to pick up at the end of the first act when they did a few early scenes from *Huckleberry Finn*. During the intermission I chatted with a local teenager behind the refreshment stand. She had worked there for several summers, sometimes onstage, sometimes off. She was friends with all the actors, many of whom went to her school.

"Who plays Jim?" I asked.

The girl looked at me quizzically. "There isn't any Jim."

The second act began with one of those wonderful conversations be-tween Huck and Jim on the raft, here transposed into idle chitchat be-tween Huck and Aunt Polly as Aunt Polly shelled peas on the front steps. No Jim. No black actors in the entire cast. And no reference to the fact that both Hannibal and the fictionalized version of it in Twain's novels had been a slaveholding town. In this "delightful" pageant, the black presence in Hannibal and in Mark Twain's work was simply erased. Yes, I had to admit as I made my way to the car in the middle of the second act, the production, just as the brochure promised, was very "special." The corn pudding began to curdle in my stomach.

The Mark Twain Outdoor Theater was upholding a long American tradition of making slavery and its legacy and blacks themselves invisible. It had been happening for more than a hundred years. A series on the Civil War that ran in *Century* magazine in the early 1890s barely mentioned slavery. Those who found it difficult to acknowledge that some of their forebears "either participated in, supported, or acquiesced in the particularly inhumane nature of American slavery," as the Southern Baptist resolution put it, would find their most useful ally in the historian Frederick Jackson Turner, who in 1893 asserted that the frontier, not slavery, was what really mattered in American history. "The slavery question" would come to be seen as a mere "incident," Turner claimed, when American history was "rightly viewed."

As I pulled into the Best Western Hotel Clemens, I thought about the other Turners that Frederick Jackson Turner had so conveniently displaced. There was Nat Turner, the leader of a slave uprising that gave the lie to the paternalistic tripe about how grateful and happy slaves were about their condition. And, closer to home, there was Missouri's own James Milton Turner, a contemporary of Mark Twain's, an ex-slave and a graduate of Meachum's "Freedom School," whose achievements as an educator and as the first black U.S. diplomat were a refutation to popular stereotypes about black intellectual inferiority. The "Reflections of Mark Twain"—and Hannibal itself, from what I had seen so far—seemed to be trying hard to inhabit the world of Frederick Jackson Turner while denying and ignoring the challenges to it that the other, more troublesome Turners posed.

* * *

"My son wasn't allowed to come to the graduation parties at Hannibal High even though he was the president of the class," said Ruth Baker evenly, as if to say that's just the way things were. She and Estel Griggsby—longtime black residents of Hannibal, both currently retired, both volunteer docents at Grant's Drug Store in the historic district—had agreed to meet me for coffee at the Mark Twain Family Restaurant Thursday morning. Ruth Baker, now seventy-seven, had worked with the mentally retarded; Estel Griggsby, now eighty-three, had done maintenance work at a local Catholic school. As we chatted about raising children, about schools, about Hannibal yesterday and today, narrative threads emerged—about race relations, about history, about Mark Twain—that would echo in conversations I would have throughout the day.

The Ku Klux Klan had been a muted but terrifying presence during her childhood, Ruth Baker recalled. Her mother had cleaned house for "a very nice woman, a lovely woman, who she was delighted to work for," but she was shocked one day when she discovered a KKK emblem while dusting. "She left and never went back." The Klan burned crosses regularly at a place called Lime Kiln Hill on the south side of town.

The Hannibal Ruth Baker and Estel Griggsby grew up in was a segregated town: restaurants, swimming pools, and other public facilities were off limits to blacks. Both attended Douglasville School. Mrs. Baker said it was named for a local black citizen and was skeptical when I said I'd heard it was named for Frederick Douglass. In any case, she said, they were never taught anything about Frederick Douglass or any other blacks in that school. They used textbooks the white schools had discarded, and no blacks ever made it into those textbooks. While both of them had heard rumors of Hannibal having been a stop on the Underground Railroad, there was never any mention of that fact in those textbooks. If whites had opposed slavery, you could be sure those textbooks wouldn't find occasion to mention it. They were surprised to hear me say that Mark Twain had spoken out against racism after the war, surprised to hear about the speech in which he said civilization had begun when slavery ended. They had always assumed Twain shared the views of the white people of Hannibal in his day. Nobody had ever suggested otherwise.

Were black and white children able to be friends in Hannibal today? I asked. They were friends until they were taught not to be, Mrs. Baker said, echoing Mayor Schwartz.

It was almost time for their shift as guides at Grant's Drug Store. As we were getting up to go, I asked Estel Griggsby where the black businesses in segregated Hannibal had been. He smiled and made a sweeping gesture. "Right here," he said, explaining that the historic district, including the area around Grant's Drug Store, used to be filled with black-owned businesses—"a barbershop, a hotel, a night club named Blue Heaven." All traces of them were gone now. That wasn't the history that the "historic district" had chosen to recover and preserve.

* * *

The Mark Twain Cave is fifty-two degrees all year round, so I took a sweater as the tourist brochure advised. I had never been in a cave before except in my imagination, so I was intrigued by every aspect of the tour. The rock formations and striations were horizontal, unlike a number of artists' renderings of the cave in *Tom Sawyer*. But the depths were as vast and intricate as any image of them I had ever seen. I recalled a critic's theory that Mark Twain was likely to have read a prominent 1874 newspaper article entitled "The Cave at Hannibal" shortly before writing *Tom Sawyer*. "A thousand weird tales are told about the cave," the writer concluded. "It is strange that some enterprising story-teller has not seized upon it as the locality of some blood and thunder novel." Twain, the critic believed, took this as "an invitation to plot a story around some boyhood knowledge he remembered about the cave."

Our guide shared some of those "thousand weird tales" as we walked past rock formations whose names were familiar to me from *Tom Sawyer*. At one point in the tour, he purposely turned out the string of electric lights that had been illuminating our path and left us to imagine the terror Tom and Becky experienced, knowing that nothing but darkness separated them from the vicious murderer Injun Joe.

Injun Joe was a major presence in Hannibal lore. I had encountered him the day before in the Haunted House on Hill Street Wax Museum. There he was considered too unsavory and dangerous to loiter with the other wax characters in Twain's books and their (also wax) real-life counterparts: he occupied a freestanding display of his own in the museum's window, while the others were arranged in a tableau inside.

I was shivering by the end of the hour-long tour. The sweater I had brought along was too light to insulate me from the damp chill. I was relieved when the tour was over and we emerged in the cave's gift shop, where we had started out. My attention was immediately caught by an item that was not for sale—a yellowed newspaper clipping framed under glass on the wall. "SAVAGE" INDIAN DISCOUNTS STORY read the headline of the undated article.

In *The Adventures of Tom Sawyer,* Mark Twain describes Indian Joe as a wild and savage Indian. His account of the Indian was discounted by Hannibal's Indian Joe, but it nonetheless made Hannibal's Indian famous. . . . The real-life Indian Joe, unlike the character described in

Mark Twain's books, described himself as a [*sic*] honest man, never harming a person and always living an honorable life.

His own story indicates he was found as an infant by a white man in an abandoned Indian camp, in Callaway County, Mo., and the man took him and reared him until he was a "good-sized" boy. When he left this place he was given a pony by his benefactor and he started out in an attempt to find out who his parents were, but was never able to do so.

From Callaway County he went to Ralls County and worked for some time for farmers near Center, always caring for the life in the wilds. He found great sport in going into the woods where he could imitate the call of almost any animal or bird. From Ralls County he came to Hannibal where he worked for various people at odd jobs, saving his money and investing it in real estate.

He bought lots in the valley between the Cruikshank home and Paris Avenue which became known as Douglassville [*sic*], an area in which Negro people built homes.

Many people claimed he was the Indian Joe of Mark Twain's writing, but this was emphatically denied by Indian Joe.

He was a unique figure in the history of Hannibal for more than three quarters of a century. He was well known and was familiar with the growth and development of the city. His exact age was not known but was believed to be about 102 years old at the time of his death.

He died on Sept. 30, 1923, and although he was married twice, he was survived by only one step-son, W. F. Johnson of Valley City.

The cause of Indian Joe's death was determined to be ptomaine poisoning from eating pickled pig's feet. He had been ill for five weeks and died at St. Elizabeth's hospital where he was a patient for one week. The home in which Indian Joe was living at the time of his death is still standing, and is located at 819 Hill St.

I stared hard at the photograph that accompanied the story. Joe Douglas was black. The "one-drop rule" that pervaded legal racial categories in the United States would have defined him as black despite the fact that he was also part Osage Indian. After Joe Douglas bought property on upper Hill Street, other blacks quickly followed. The area—Douglasville, named after Joe Douglas—soon became the center of Hannibal's black community.

How did Joe Douglas become the model for a base and murderous figure like Injun Joe? I soon realized that I was asking the wrong question. The real question was why the good people of Hannibal had *decided* he was the model. Mark Twain never said Joe Douglas was Injun Joe. No, it was the citizens of Hannibal, giddy with the challenge of pairing each fictional creation of Twain's with a real-life counterpart, who had made that match. As a result, a man who should have been honored as hardworking and enterprising was unjustly remembered as the model for a murderous villain in the most famous novel by the town's most famous native son.

I remembered a 1902 article I had seen yesterday in the Mark Twain Museum on Hannibal's propensity to come up with real-life matches for Twain's fictional characters.

> You don't need to bait your hook if you are fishing for a Huck. Just make a cast anywhere around town and there's your Huckleberry. "I've lived in this town 71 years," remarked one reputable native, "and I have never been accused of being Huckleberry Finn until last night. I submit that my reputation has been good, and if I ever was Huck Finn, I can't recall it. It's a base slander to put upon a man in his old age." Each man, woman and child in Hannibal seems to have agreed upon a different Huck, Tom and Becky. . . .

Be that as it may, only one local resident was ever tagged as Twain's Injun Joe. Joe Douglas may have been known to some whites in Hannibal by the nickname "Indian Joe," but there any resemblance between him and Twain's treacherous, violent character abruptly ends. Although sources diverge on various particulars, there are some basic consistencies in the stories about Joe Douglas: he was known as honest and resourceful, and he was a storyteller who entertained the local children.

Mark Twain narrates the death of Injun Joe on two separate occasions. First he starves him in the cave in *Tom Sawyer* and then, implying that he concocted that earlier scene simply "to meet the exigencies of romantic literature," he describes another death in his autobiography, where he imagines Satan coming to snatch Injun Joe's soul in a fierce thunderstorm. But while Twain relishes the opportunity to tell us how Injun Joe died, he is absolutely mute about how Joe Douglas died. There is at least one good reason for that: Joe Douglas outlived Mark Twain by thirteen years. The clipping at the cave said Douglas was raised by a white man; another source said he was raised by a black family; but most agree that Joe

Douglas died in 1923 at age 102, either of ptomaine poisoning or from pneumonia.

According to an undated article, probably from 1894,

> Joe Douglas, good-humoredly called "Injun Joe," out at Hannibal, Mo., for one of the sinister characters in Mark Twain's book, Tom Sawyer, resents the nickname, declaring his life has not warranted such a character as that given to Injun Joe in the book. And his friends in Hannibal indorse [*sic*] the statement.

This article corroborates much of the information in the clipping at the cave, noting that in contrast to the Injun Joe of *Tom Sawyer*, Joe Douglas "worked hard, and saved his money and prospered. The people about Hannibal like him, and when they refer to him as 'Injun Joe' it is with no thought of anything bad about him." The article goes on to cite the reminiscences of a man who would become Hannibal's mayor (and whom Twain would recall in his autobiographical dictations): "Doc Buck Brown, of Hannibal, who was a boyhood chum of the creator of Tom Sawyer, declares that the real Injun Joe, whom he knew well, was anything but a bad man." His interviewer, however, remained incredulous:

> "But didn't he kill young Doc Robinson?—that's what it says in the book."
>
> "Tut, tut," returned Doc Brown, "Injun Joe never killed anybody. He was as peaceable as a kitten—showed the boys the good swimming holes and helped 'em to find clam shells, and cooked fish for 'em."

Even in Twain's own writings there are two Injun Joes. One is the degenerate villain of *Tom Sawyer*. The other is a harmless local personality who occasionally drank too much and told the boys stories. Twain wrote that his father "once tried to reform" this Injun Joe.

> It was a failure, and we boys were glad. For Injun Joe, drunk, was interesting and a benefaction to us, but Injun Joe, sober, was a dreary spectacle. We watched my father's experiments upon him with a good deal of anxiety, but it came out all right and we were satisfied. Injun Joe got drunk oftener than before, and became intolerably interesting.

In 1952 Dixon Wecter raised the notion that the "drunkard and murderer" in *Tom Sawyer* "is probably touched up a good deal from reality"—if, indeed, he is based on "reality" at all. Clearly the "exigencies of romantic literature" rather than any effort at verisimilitude shaped Twain's melodramatic villain. The "real Injun Joe," Wecter notes, describing a figure whose early history somewhat resembles that of Joe Douglas,

> earned a little cash by toting carpetbags between wharf and tavern . . . claimed to remember Sam Clemens, and died in his nineties a respected citizen. His villainy, too, may have been largely imaginary, his vindictiveness perhaps a memory from the days of Murrell's Gang, whose bloody acts were the common tradition of all river towns in Sam's boyhood.

Later, in a local bookstore, I browsed through a book called *Hannibal Yesterdays* by Hurley and Roberta Hagood, where I learned that

> In 1870, the population of Hannibal was 10,125 and of this number 1,616 were Negroes. In 1870, a three room school for Negroes was built by the Hannibal Public School System on Rock Street between Ninth and Tenth Streets in a community called Douglasville. The school was named Douglasville School, and was built only after many requests for it had been made to the Hannibal school board. . . .
>
> The Douglasville community had been named for Joe Douglas, a part Indian man who lived in the 1000 block on Hill Street, and who acquired property in the area by saving his meager earnings. The Douglasville School had taken its name from its location, named for Joe Douglas.

History had played a mean trick on Joe Douglas, the namesake in 1870 of Hannibal's first black school. Tom Sawyer's gang had pulled off an act of thievery for the books, all right. They had managed to rob Joe Douglas of his good name and saddle him with the reputation of having been the model for one of the most reprehensible miscreants in American literature.

A curious headstone stands about a hundred yards from where Sam Clemens' parents, sister, and brothers are buried in Hannibal's Mt. Olivet Cemetery. It reads as follows:

INJUN JOE

> Joe Douglass, known to many in Hannibal as Injun Joe, died September 29th, 1923 at age 102. He was found, an infant, in an abandoned indian camp by a man named Douglass who raised him. He denied that he was the Injun Joe in Mark Twain's writings, as he had always lived an honorable life. He was buried from A.M.E. [African Methodist Episcopal] Church.

4. *The "Injun Joe" headstone on Joe Douglas's grave in Mt. Olivet Cemetery in Hannibal, Missouri (photo credit: Shelley Fisher Fishkin)*

How did that strange headstone get there—and with Douglas's name spelled wrong, no less? Surely his family and friends would have fought to prevent this effort to link the poor man after death with the legend that had tormented him during his life.

Hurley Hagood told me the story. The stone, a fairly recent addition, was about ten years old. It had been the brainchild of Eugene W. Yarbrough, the enterprising former caretaker of Mt. Olivet Cemetery. When Yarbrough took over as caretaker, the cemetery was in poor shape physically. A hard worker, he fixed up the roads and made the entrance more attractive. Once the cemetery was in better condition, more people began to come. But as a tourist attraction it was lacking: everyone wanted to know where Injun Joe was buried. Yarbrough knew where Joe Douglas's grave was at Mt. Olivet and arranged for the new headstone (a relative of his was the stonecutter), complete with the inscription and the misspelled name. (He may have been thrown by the fact that the school once named for Joe Douglas had been renamed for Frederick Douglass; but had he glanced at the adjacent headstone of Douglas's wife, on which the family name is spelled correctly, he might have avoided the error.) While the inscription's reference to his "honorable life" might have pleased Joe Douglas, the cost—having INJUN JOE emblazoned on his tombstone in oversized capital letters—might well have struck him as too high. ("Joe Douglas, Real Estate Developer" might have suited him just fine.) His wife's grave bears a simple inscription: "Anna S. Douglas, Wife of Joe Douglas, born Lynchburg, Va., died Feb. 17, 1902 age 52, Rest in Peace."

It must have been hard for Joe Douglas to lose a wife twenty-nine years his junior. When Mark Twain came to town a few months after Anna Douglas died, on what was to be his last visit to Hannibal, he gave a speech from the steps of Rockcliffe Mansion, just down the street from where Joe Douglas lived. Did his presence open old wounds, remind Joe Douglas of the blasted book that had required him to spend the rest of his life explaining himself, defending himself, denying outrageous innuendoes to strangers? Or did it recall old memories—of showing the boys the best swimming holes, of telling them stories, of having them ask for more? We'll never know. Joe Douglas left no written record. His house in Douglasville was there until 1994, when it was unceremoniously torn down as a hazard to children.

* * *

Dixie M. Forte carefully turned the fragile pages of the yellowing Hannibal *Colored Directory*, her eyes scanning the columns of names, occupations, and addresses.

"Mrs. Lula Clay? Lula Clay was a hairdresser? Now I remember her. She was an agent for Madame C. J. Walker. And here's Dr. W. C. Conway—he had his own dentist office. Here's Archy Corsey, he was a deliveryman. It's got everything that they did."

"Here's my husband, James Henry Crow, employed at the cement plant," interjected Hiawatha Crow, a former teacher and member of the Hannibal City Council.

"Whoever kept this book kept it in really good shape," said Mrs. Forte. She had been offered "a lot of money for it" when she had her antique shop in town, but she wouldn't sell it. It was a piece of history she planned to keep. She had brought it to our lunch at the Mark Twain Family Restaurant because she knew of my interest in Hannibal history. Mrs. Crow had brought a large scrapbook of clippings.

Reverend Ann Facen, pastor of the Willow Street Christian Church, Disciples of Christ, arrived and made our luncheon party complete. I had called these women—all longtime Hannibal residents and leaders in the black community—at Mayor Schwartz's suggestion, and they had agreed to meet me for lunch on Friday afternoon. Each of us ordered Mark Twain Fried Chicken, the restaurant's specialty.

As Mrs. Forte closed the Colored Directory, she pointed out a paragraph on its first page about Hannibal being Mark Twain's town.

"You know, when I was on a mission tour in England, I asked some young people where they'd like to go if they came to the United States," Reverend Facen said. "They surprised me and said Mark Twain country!"

Reverend Facen hadn't paid much attention to Mark Twain until she moved to Hannibal ten years ago. Walking by the local park with her daughter, she came across a historical marker that referred to "Nigger Jim." The two of them were incensed. When they got home, Reverend Facen called the NAACP. They petitioned successfully to have the racial slur removed. (Indeed, one of Mayor Schwartz's first acts in office, he had told me, was taking care of that problem.) "At first they just painted over the word 'Nigger.' Then they took the marker down." (Twain himself, incidentally, never calls his character "Nigger Jim." Critics and writers—including Albert Bigelow Paine, his first biographer, and Ernest Hemingway—ascribed the phrase to him and etched it into the popular imagination.)

I mentioned that a dimension that seemed to be missing from the local commemoration of Twain was how strongly he attacked racism after he left Hannibal.

"You never hear about that," Reverend Facen said. "It's not brought out."

I told them about a conversation I had had with Roberta and Hurley Hagood about responses to their research on Hannibal's Underground Railroad. They got letters from people saying, "Why weren't we taught this in school?" I said it seemed to me that the stories of blacks and whites who worked to create a different kind of society were empowering; if those stories got erased, then we were transmitting a very strange version of the past. "How do you account for the version of history that gets passed on?" I asked.

"Who were the printers and the writers?" Reverend Facen asked, suggesting that those who wielded the power of the pen saw no need to highlight the often heroic stories of blacks and whites who worked for social and political change. "It's not lifted up. Even in the midst of that struggle, you see strength and determination of a people who really endured. But that's not lifted up. And it's intentional. Some of it is by chance; most of it, I feel, is really by design." These omissions saddened her, she said, because children needed to hear these stories that spoke to problems plaguing their lives today.

"Right now in Hannibal this past year out at the school," she continued, a gang that called themselves the Cowboys took to "carrying the Confederate flag in their pockets" as part of "the 'hate nigger' thing." It was as if "the sixties never happened," she said. (Later a teacher would tell me about the jackets a few students were in the habit of wearing at Hannibal High that said "H.A.N.K." on the back: "Honkies Against Niggers Klub.")

Reverend Facen's grandson, who was in the eighth grade, had been living with her for the year, so she had had direct contact with what was going on. The tensions generated by the Cowboys' bravado began to erupt in incidents. Black children were told not to walk down the halls together. "Teachers would say, 'Break it up.' But white kids could be in clusters." (It sounded oddly reminiscent of a law the Missouri legislature had passed in 1847 forbidding groups of blacks to assemble.) Last summer some Cowboys defaced the walls of the playground at Reverend Facen's church with graffiti. "Cowboys were riding in trucks, pickups, saying that if any of the white kids were friends to the black kids, they were . . . 'nigger lovers.' "

Reverend Facen's grandson, who had just turned fourteen, told her of feeling shunned by the white students and ignored by the white teachers. (There are no black teachers in the middle school or the high school.) The white student seated next to him at band practice made no secret of his distaste for the fact that they were supposed to share a music stand. When Reverend Facen's grandson made first chair in the trumpet section, the band teacher "added the score three or four times. He couldn't *believe* it," she said. The boy he beat was subjected to taunts from his white classmates in the halls: "Ooh. You let *him* beat you?" Black students were further demoralized, she maintained, by a double standard of justice: when white children got into fights, they got a detention; when black children got into fights, they got expelled.

Reverend Facen had taken action. She involved the NAACP and the superintendent of schools, and was trying to start a group of black and white parents "because we cannot have this," she said. "You would think we're beyond that. But we're just now beginning to sit down and talk and listen to each other." While she found the failure of the schools to deal with racial tensions disturbing, she was even more upset by the inaction of the churches. In the past there hadn't even been a multiracial ministerial fellowship. "They just told me, 'Hannibal is just that way.' We couldn't even come together to pray!"

She paused. "As for Mark Twain, and how he addressed racism? You *never* hear about that. It's sad."

I told them the story about "Black John," the teenage slave who was the best friend of Tom Blankenship, a boy on whom Twain consciously modeled aspects of Huckleberry Finn. According to one of Twain's childhood friends, "Black John and Tom Blankenship were naturally leading spirits and they led us younger 'weaker' ones through all our sports. Both were 'talented,' bold, kind, and just and we all 'liked' them both and were easily led by them." So Twain was following a black kid, a natural leader, all during his childhood while he was running around the Hill—not a bad model for how black and white fourteen-year-olds can get along with each other. "Why don't they pass that story along? It might help," I suggested.

"It probably would," said Reverend Facen. "But they don't have it in there."

"Has there ever been any move to create a black history museum in Hannibal?" I asked.

"They had something at the library two or three years ago," Mrs. Forte answered. Reverend Facen said her church had bought the property that had been the black school. They had tried to interest the historical society in preserving and restoring the building and making it a museum for black history, but it didn't happen: no funds.

But there had been some progress, the group agreed. Martin Luther King Day was a holiday in the schools. The black community had celebrations all over town, and most churches sponsored weeklong activities culminating in a march. There was more awareness, they said, than there had been years ago.

I mentioned the young black man named Jerry, the Missouri slave Mark Twain remembered fifty years later as a consummate orator, someone who had impressed him as a child as "the greatest man in the country." I also brought up the important impact Mark Twain had had on black writers, including Ralph Ellison and Langston Hughes. These were the kinds of things, said Dixie Forte, that would have been nice to know about.

A broad swath of history was pasted into the large brown scrapbook Hiawatha Crow pulled out and placed on the table. We all crowded around and browsed through the newspaper clippings it contained. The first clippings we came across dealt with Ku Klux Klan organizing efforts in Hannibal in 1982. KKK RECRUITERS EXPECT "GOOD RE-SPONSE" HERE read the headline of a story in the local paper.

The thirty knights of the Ku Klux Klan are now working in the Hannibal area to recruit new members, said a St. Louis-based organizer for the KKK . . . [who] added that the KKK has sympathizers and secret members holding political office in Northeast Missouri. . . . "The U.S. Constitution was created for the white man. All these other creatures—it wasn't meant for them," she said. . . .

The next clipping, from the *Boston Herald American*, bore the headline IGNORE THE KLAN. It read:

That the Klan seeks to sully the integrity of Hannibal is bad enough; but Hannibal is a significant place for all Americans because of Samuel Clemens. All of Sam Clemens' writings were rooted in the values he knew from growing up in Florida and Hannibal.

The Klan wants to rally in the same town Mark Twain used in *Huckleberry Finn*. It is Huck who befriends and saves the escaped slave, Jim. The white boy and the black man camped out on the island that can still be visited across the Mississippi from Hannibal, a bold thing to write in 1885. From the looks of things today, this human brotherhood is still being threatened. Mark Twain ran against the current of his time. His fellow Americans always disappointed him as they disappoint me. . . .

The lunch crowd had long since departed from the Mark Twain Family Restaurant. We had made it through only the first few pages of Hiawatha Crow's dense scrapbook; somehow the afternoon had gotten away from us. "You know, I've got more in the car," Mrs. Crow said. We headed out back, and sure enough, there was a carton full of scrapbooks, each as packed as the one we had been looking at. I took a few minutes to leaf through the one on top. There were clippings from black newspapers from the twenties and thirties, stories about Frederick Douglass's daughter, about Mary McLeod Bethune, about Harry Truman. Harry Truman? I paused to find out what this son of Missouri had done to warrant a spot in Mrs. Crow's scrapbook. The AP story that ran in 1983 in the Hannibal *Courier-Post*, noting the publication of a volume of Truman's letters, included the following:

Although he gained a reputation for advancement of civil rights as President, in a letter dated June 22, 1911 the twenty-seven-year-old farmer Truman shared his racial views with his future wife. "I think one man is just as good as another so long as he's honest and decent and not a nigger or a chinaman," he wrote.

This from the man who signed the executive order integrating the U.S. armed forces! We shook our heads. I could have spent a week or more in those boxes in the back of Mrs. Crow's car, but we all had appointments to keep. Reluctantly we said our good-byes.

*   *   *

Traveling north toward Palmyra on Highway 61, I passed the Injun Joe Campgrounds and the Huck Finn Shopping Center. I turned north to 168, then north again on Rural Route JJ. Lush green trees lined the road as rolling hills leveled off into flat, open fields. I crossed the railroad tracks near a sign that said "South River." I passed a fertilizer plant on my left and a power plant on my right. The road dead-ended at the entrance to the American Cyanamid plant, just as Hurley Hagood said it would when he wrote the directions down for me. I stopped the car. "Private Property. No Trespassing. Violators Will Be Prosecuted." Twain's "Notice" at the start of *Huckleberry Finn* came to mind. Would anyone looking for a moral be banished, I wondered, or anyone looking for a plot be shot? I got out and tried to imagine what this place had been like a century and a half ago.

It was here, on July 12, 1841, that two theology students from the Mission Institute in Quincy and a mechanic who had long been active on the Underground Railroad crossed the Mississippi from Illinois by canoe and came ashore between Hannibal and Palmyra to help some Missouri slaves gain their freedom. Before the day was done, the three men from Illinois had lost their own. George Thompson, James Burr, and Alanson Work were sent to the Missouri Penitentiary by a jury that included John Marshall Clemens.

I recalled Twain's comment in his autobiography that in his "schoolboy days [he] was not aware that there was anything wrong about [slavery]. No one arraigned it in my hearing; the local papers said nothing against it; the local pulpit taught us that God approved it." Yet elsewhere in that same book he casually observed that his brother Orion was "an abolitionist from his boyhood to his death." How did Clemens' brother— "born and reared among slave-holders," as Clemens was—get to be an abolitionist in such a monolithic environment? The drive I had taken with Hurley Hagood the day before, the articles and letters he and his wife had shared with me, and the research I had pursued prior to my trip had taught me that there were more challenges to slavery in the Hannibal of Mark Twain's childhood than he had chosen to recall in his autobiography. Ten years older than Sam, Orion Clemens seems to have been more attentive to these undercurrents than his brother was.

"Go straight ahead," Hurley Hagood had directed as I drove up Harrison Hill. "You see that place across over there . . . where the sign is, and then there's a house?" Like the other places Hurley Hagood had taken me, the house around the bend, across from the sign, up the hill on the corner of Harrison Hill and Driftway Drive wasn't distinguished by any historical marker, hadn't made it into any historical register, didn't appear in any brochure put out by the Hannibal Visitors and Convention Bureau, and wasn't a stop on the Twainland Express. There had once been many outbuildings and cabins on the grounds that were used to hide slaves during daylight hours; the escapees would leave for the river after dark to cross at designated sites, guided by "conductors." When the house was remodeled, workmen removed a slate wall that had lists of names scratched on it—the names of slaves who had passed through.

Runaway slaves were hidden in a variety of places throughout Hannibal during Mark Twain's childhood, one of them reputedly the cave he immortalized in *Tom Sawyer,* on the south side of town where few slaveholders lived. Secret hiding places were said to have existed in a house on Spring Street connected by underground tunnel in the basement to a spring house, in a two-story brick double house on Birch Street, in a house on Valley Street near the kilns of the Hannibal Lime Company. This behind-the-scenes activity, however, was highly guarded and secret, for as Twain wrote in 1894 in a story that wasn't published until after his death, "In that day for a man to speak out openly and proclaim himself an enemy of negro slavery was simply to proclaim himself a madman. For he was blaspheming against the holiest thing known to a Missourian, and could *not* be in his right mind."

If those who spoke out against slavery were thought mad, those who actively interfered with it were considered dangerous. It mattered not that John A. Lennon—a man described by Orion Clemens in the *Daily Journal* as "clever, witty and entertaining . . . an excellent speaker"—had taken a leading part in Hannibal's civic and religious affairs; when his ties to the Underground Railroad in Quincy, Illinois, became known in 1855, he and his family were run out of town by Archibald RoBards, a former mayor of Hannibal and the father of one of Sam Clemens' classmates. The Quincy route was the most widely utilized one from Hannibal, although sometimes "passengers" were taken to the Iowa border. From Quincy the slaves went to Galesburg or Springfield and on to Chicago and Canada. Links to Quincy had been strong since 1835, when Dr. David Nelson, the abolitionist president of Marion College, near Hannibal, fled there after a delegation of local proslavery farmers ordered him to leave town. In Quincy he became associated with Eels College, later known as the Mission Institute, where Thompson, Burr, and Work were based when they made the decision to cross over to Missouri that July day in 1841.

According to a letter Alanson Work sent from prison, as soon as the three young men came ashore near this spot, the present site of American Cyanamid, they saw "a woman and her son hoeing tobacco." Speaking to her, they "found that she wanted to be free and agreed to help her." They had similar conversations with the other slaves and arranged to meet them at an appointed place after dark. The slaves appeared, as promised, but as their would-be liberators "started in a footpath for the river, rejoicing in the prospect of helping the oppressed to liberty and happiness," the slaves turned on them, tied their hands, and betrayed them to their masters. The next morning the prisoners were bound to a sixteen-foot chain and marched to the courthouse in Palmyra. (Why had the slaves betrayed them? Not loyalty, as their masters boasted, but fear was the most likely motive. One of the masters testified under oath that he and two friends overheard Work and Burr in conversation with "my negroes." The slaves knew escape was impossible and were pressured and bribed by their masters to turn the abolitionists in.) In keeping with the widely held view that an open opponent of slavery was a self-proclaimed "madman," the local paper "lamented that Missouri had no 'Lunatic Asylum,' as 'the poor, deluded creatures' were victims of 'monomania—a case not where the morals are stained, but where the mind is disordered.' "

While in jail in Palmyra, the accused were subjected to a constant barrage of abuse. There was, Thompson writes, "a tremendous excitement all over the country against us. Even the *little boys* drank in the spirit, and would come to the jail and try to torment us, knocking on the door and calling out,—'Ha! there, nigger-stealers, you think you will steal any our niggers, heh?' " Well-dressed women similarly had no hesitation about stopping to curse them at every opportunity. Burr, Thompson, and Work were accused of grand larceny since stealing property was the only crime they could be legally charged with and each slave was worth approximately three hundred dollars. The men protested that they had no

THE INTERIOR OF THE JAIL IN PALMYRA, MISSOURI.

WE were confined in this place eleven weeks, previous to our being taken to the
Penitentiary, at Jefferson City. SEE PAGE 18.

*5. Interior of jail cell in Palmyra, Missouri, picturing Underground Railroad
operatives Alanson Work, James E. Burr, George Thompson, and other prisoners
(The sketch appeared in Thompson's* Prison Life and Reflections, *Hartford, 1851)*

interest in stealing property; their only interest was in helping the "property" steal *themselves* to freedom. ("The slaves own themselves by the law
of God," Thompson wrote in a letter to a friend, prefiguring Jim's comment in *Huckleberry Finn*, "I owns mysef.") The able defense attorney,
Samuel Taylor Glover, made it very plain that it was absurd to charge
his clients with grand larceny, but to no avail. (Indeed, Twain would
later recall that the town always said Glover "was a fool and nothing *to*
him." Could Glover have served as a model, in part, for "Pudd'nhead"
Wilson, another lawyer marked by his community as a fool?) Thompson
reported that the jurors' "hearts were so filled with deep-rooted preju-
dices, and their minds so blinded to everything that does not uphold
slavery" that they "appeared to see no force in his reasoning." The jury
quickly returned a verdict of "Guilty and Twelve Years in the Penitentiary."

Burr and Thompson may have been relatively new at guiding slaves
safely to freedom, but Alanson Work was a pro. Known for having main-

tained a busy station on the Underground Railroad, he was said to have helped some four thousand slaves reach freedom before his prison term ended his career. There is no way of knowing how many more he would have helped had he not been taken out of service by that Palmyra jury.

John Marshall Clemens was proud of his service on the jury at this dramatic trial. His exercise of civic duty was a rare experience in competency for a man who had lurched from one personal economic disaster to another. But if John Clemens had taken pride in his role in the trial, his son Sam, years later, would just as soon have forgotten it. When Missouri historian R. I. Holcombe wrote Twain in Hartford in 1883 requesting family recollections about the trial for a history he was writing of Marion County, Twain palmed the letter off on Orion.

I took a last look at the flat fields fringed by trees and noticed how exposed it all was, how open. Then I snapped some pictures and got into the car. As I made my way back to the main road, I thought about Alanson Work and John Marshall Clemens in that courtroom in Palmyra, one in the defendant's stand and the other in the jury box, both about to leave the world remarkable legacies that neither would have had any reason, at that time, to suspect. Waiting on the Illinois shore for his father's release from prison was Work's son, a boy approximately Sam Clemens' age. Both boys became printers in their adolescence. But it was not as printers that each would leave an indelible mark on American culture.

Henry Clay Work became a songwriter and composer, and was responsible for some of the most rousing songs of the Civil War, including the acclaimed and notorious "Marching Through Georgia." After the war he wrote such favorites as "My Grandfather's Clock" as well as many popular temperance songs. He made his home in Connecticut, having moved there with his mother while his father was imprisoned in Missouri. (After receiving a pardon from the governor some years later, with the condition that he permanently leave Missouri, Alanson Work joined his family in Connecticut.) When Henry died suddenly of heart disease at age fifty-two in Hartford in 1885, the *Courant* noted in his obituary, which included a description of his father's activities on the Underground Railroad,

Our country has produced few song writers whose works have been more widely sung than Mr. Work. Some few of his productions have not only been on the lips of nearly every man, woman, and child in America, but have been known and sung, with some variations, in every part of the world. This can be said of his "Marching Through Georgia," a song which was known to every Union soldier in the war, the music of which has since appealed to every clime.

Living in Hartford himself at the time, Twain was likely to have seen Work's prominent obituary in the newspaper he read every day. The request for information from the Marion County historian two years earlier would have given Twain a recent reminder of Alanson Work's name (the historian's book came out in 1884, and a copy was in Twain's library at his death). A few years after Henry Clay Work's death, the sculptor Karl Gerhardt wrote Twain, "The author of 'Marching through Georgia' surely deserves a statue." He hoped Twain might be interested in paying for it. Twain's response is not known.

The year Henry Clay Work died in Hartford, Mark Twain published *Huckleberry Finn*. I thought about *Huckleberry Finn* as I retraced my route back to Hannibal. I thought about how Huck berated himself for helping Jim.

> I tried to make it out to myself that I warn't to blame, because I didn't run Jim off from his rightful owner; but it warn't no use, conscience up and says, every time, "But you knowed he was running for his freedom, and you could a paddled ashore and told somebody." That was so—I couldn't get around that, noway.

And I thought about how Huck told the smallpox lie to protect Jim and then "got aboard the raft feeling bad and low, because I knowed very well I had done wrong, and I seed it warn't no use for me to try to learn to do right; a body that don't get started right when he's little ain't got no show." I remembered the fear Huck expressed later in the book.

> It would get all around, that Huck Finn helped a nigger to get his freedom; and if I was ever to see anybody from that town again, I'd be ready to get down and lick his boots for shame. That's just the way: a person does a low-down thing, and then he don't want to take no consequences of it. Thinks as long as he can hide it, it ain't no disgrace. That was my fix exactly. The more I studied about this, the more my conscience went to grinding me, and the more wicked, and low-down and ornery I got to feeling. . . . I tried the best I could to kinder soften it up somehow for myself, by saying I was brung up wicked, and so I warn't so much to blame, but something inside of me kept saying, "There was the Sunday School, you could a gone to

it; and if you'd a done it they'd a learnt you, there, that people that acts as I'd been acting about that nigger goes to everlasting fire."

And I recalled Huck's response after he tries his hardest to do "right" by writing the note to Miss Watson: "your runaway nigger Jim is down here two mile below Pikesville." First he feels "good and all washed clean of sin for the first time" in his life. But then he

> got to thinking over our trip down the river; and I see Jim before me, all the time, in the day, and in the night-time, sometimes moonlight, sometimes storms, and we floating along, talking, and singing, and laughing. But somehow I couldn't seem to strike no places to harden me against him, but only the other kind. I'd seen him standing my watch on top of his'n, stead of calling me—so I could go on sleeping, and see him how glad he was when I come back out of the fog, and when I come to him again in the swamp, up there where the feud was; and such-like times; and would always call me honey, and pet me, and do everything he could think of for me, and how good he always was, and at last I struck the time I saved him by telling the men we had small-pox aboard, and he was so grateful and said I was the best friend old Jim ever had in the world, and the *only* one he's got now, and then I happened to look around, and see that paper.
>
> It was a close place. I took it up, and held it in my hand. I was a trembling, because I'd got to decide, forever, betwixt two things, and I knowed it. I studied a minute, sort of holding my breath, and then says to myself:
>
> "All right, then, I'll *go* to hell"—and tore it up.

How the hell did a little boy named Sam Clemens, who grew up in a slaveholding town, in a slaveholding family, with a father who sent "slave-stealers" to jail, end up writing a book like *that?* Then I realized that once again I'd been asking the wrong question. The right question, I decided, was how could anyone *but* someone who had grown up in a slaveholding town, in a slaveholding family, with a father who sent "slave-stealers" to jail end up writing a book like *that?*

Mark Twain became a brilliant chronicler of the process by which people who thought of themselves as good managed to tolerate evil in their midst. In *A Connecticut Yankee,* for example, when people respond to the brutal whipping of a slave by commenting "on the expert way in which the whip was handled," Twain writes,

They were too much hardened by lifelong everyday familiarity with slavery to notice that there was anything else in the exhibition that invited comment. This was what slavery could do, in the way of ossifying what one may call the superior lobe of human feeling; for these pilgrims were kind-hearted people, and they would not have allowed that man to treat a horse like that.

Elsewhere he says,

It is commonly believed that an infallible effect of slavery was to make such as lived in its midst hard-hearted. I think it had no such effect—speaking in general terms. I think it stupefied everybody's humanity, as regarded the slave, but stopped there. There were no hard-hearted people in our town—I mean there were no more than would be found in any other town of the same size in any other country; and in my experience hard-hearted people are very rare everywhere.

The challenge for Mark Twain as a writer was to reach that deadened lobe of feeling in his readers and awaken them to the humanity they shared with those whom their society wrote off as less than human. Who could be better up to that challenge than someone whose own feelings had been "ossified" but who had later come to recognize that fact and be ashamed of it?

The Hannibal of Mark Twain's youth was permeated by what Forrest Robinson has called "bad faith": the unthinking hypocrisy of people who daily violated the moral norms to which they paid lip service while pretending they were doing nothing of the sort. Twain eventually came to understand the phenomenon and the self-delusion and communal collusion required to sustain it. By dissecting the human relations that he obtained in the world of his childhood, Twain would become an acerbic analyst not only of "bad faith" but of its close cousin, the "lie of silent assertion" that could hold an entire nation in its thrall. "It would not be possible," he wrote in 1899,

for a humane and intelligent person to invent a rational excuse for slavery; yet you will remember that in the early days of the emancipation agitation in the North the agitators got but small help or countenance from any one. Argue and plead and pray as they might, they could not break the universal stillness that reigned, from pulpit and press all the way down to the bottom of society—the clammy stillness created and maintained by the lie of silent assertion—the silent assertion that there wasn't anything going on in which humane and intelligent people were interested.

Most residents of Hannibal during Twain's childhood clearly bought the "silent assertion that nothing is going on which fair and intelligent men are aware of and are engaged by their duty to try to stop." Huck had bought it as well, and out of Twain's probing of the implications of that fact came a stunningly powerful novel.

The erasure of the black presence in Mark Twain's world angered and disturbed me, as did the erasure of the history of black and white resistance to slavery and racism. If we cut this part of history out of the picture, we lose stories of courage and spirit, of survival and defiance and struggle. And we lose a sense of the complex forces which shaped both the work of Mark Twain and the work of the nation. The result is a diminished view of both. A whitewashed fence is one thing, I thought. A whitewashed history is another.

It was the idyll of childhood that was being promoted in Hannibal, nostalgia for a simpler world. But it was nostalgia for a world that never was—not even in Twain's books. Even in his earliest literary renderings of Hannibal, some of the town's darker dimensions break through. At the center of *Tom Sawyer* is a preoccupation with the bad faith and hypocrisy that pervade adult society; *Life on the Mississippi* and *Huckleberry Finn* continue the exposé of what tried to pass itself off as "civilization" in the South. As an artist, Twain returned to the Hannibal of his youth again and again, training on it the complex double vision that recreated both what he saw as a child and what he saw as an adult. The underside of the idyll, largely beneath the surface in the earliest work, would become more prominent as Twain matured. Like the man that corrupted Hadleyburg, Twain held those hypocrisies and pretensions up to the light for all the world to see.

Hannibal—the lies it told itself, the truths it denied—would serve as an emblem for Twain of the larger "civilized" world. Twain's Hannibal, in turn, became an emblem of America itself. To the extent that one could mask the betrayals and depravities at the heart of Hannibal, one could deny the dark underside of the nation.

Clearly Hannibal was not alone in avoiding chapters of the past that were still painful and difficult. It was a problem the whole country shared. I recalled my visit, a few months earlier, to my friend Jim Horton, a professor of American history at George Washington University, who had been on leave working for the National Park Service as special assistant to the secretary of the interior. He had shown me the plans that were in the works for a project that would commemorate Underground Railroad sites in a guided drive from some point in the Deep South to a destination in the North, with historical markers at key places and perhaps also tape recordings one could listen to in the car as one drove from site to site, dramatizing the danger-fraught and heroic journeys that had taken place. And I had heard talk about an Underground Railroad museum being planned in Cincinnati. As a nation, we were just beginning to put our historical house in order. And the record of nineteenth-century black life that I had found so tantalizing to track through Mark Twain's writing and so shockingly absent during my trip to Hannibal? Only a handful of museums and historic sites nationwide did justice to this part of American history.

In the spring of 1996, when I was writing this book, an exhibit entitled "The Cultural Landscape of the Plantation," originally scheduled for the Library of Congress, opened at the Martin Luther King Jr. branch of the Washington, D.C., public library. It used photographs and the words of slaves to show the world of the plantation from the slaves' point of view. "The experience of slavery and the institutional racism that still pervades American society had its beginnings here," said the curator, Professor John Vlach, in an interview on PBS. But the sight of an armed white overseer in a field of slaves picking cotton had so enraged black employees of the Library of Congress that they protested against the exhibit and demanded its removal. I thought of the perennial efforts to remove *Huckleberry Finn* for similar reasons: works like novels and photographs that give a human face to the racism of the past and its legacy in the present are troubling, disturbing, dangerous. How much easier simply to expel them from view.

*   *   *

"See this famous river town from the boat that bears the name of its most famous citizen . . . Mark Twain," read the 1995 Visitor's Guide to Hannibal. "Enjoy one-hour Sightseeing Cruises with commentary on river history, legends and sights. See Jackson's Island, riverboats and barges, Lover's Leap and the mighty Mississippi River rolling along. Beverages of your choice and sandwiches available at the snack bar. (No reservations needed)."

There was one last stop I had to make before leaving town. I had been on riverboats before, in New Orleans and San Diego, both times under the auspices of the Mark Twain Circle of America, with a boatload of scholars talking Twain talk for hours while cruising the Mississippi or circling Mission Bay. No, it wasn't the novelty of riding a steamboat up the river that got me aboard the *Mark Twain* in Hannibal. I had signed on for the tour for one reason: I wanted to get close to Jackson's Island.

Its real name had been Glasscock's Island, but like other places in town that were known by the name Mark Twain gave them, it was Jackson's Island now. In real life Bence Blankenship had smuggled food to a runaway slave on one such island that dotted the river between Hannibal and the Illinois shore. But in that realm that seemed to have greater reality than life itself, Jackson's Island was the scene of Huck Finn's first encounter with Jim after Jim fled from Miss Watson's. It was where their friendship was forged. It was where Huck raced to warn Jim about the approaching slave-catchers, shouting, "They're after us!" thereby casting his lot with the runaway slave who became the closest thing to a father he would ever know.

The whistle sounded and the *Mark Twain* pulled out of the Center Street landing. It passed under the Mark Twain Bridge, a link in Highway 36, one of the major east–west routes in the country between Indianapolis and Denver. I had driven over the bridge earlier, curious about how long it took to make that trip from Missouri to Illinois today: one minute. In Twain's childhood it might have taken a slave many weeks of planning to make that journey and still face great danger after reaching Illinois, where slave catchers seeking rewards and bounties combed the shore.

The guide drew our attention to the Mark Twain Lighthouse on Cardiff Hill. The bright afternoon sun glinted off the metal beams of the bridge and made the white lighthouse stand out against the lush greenery that surrounded it. Although they both had their failings—the Mark Twain Bridge was about to be torn down and replaced by a larger, stronger structure, and Mark Twain Lighthouse had a tendency to go out in storms (it had to be relit and rededicated by Presidents Kennedy and Clinton)— each had a symbolic resonance that appealed to me. I liked the idea of a Mark Twain Bridge joining a former slave state to a free state, and linking the country from east to west, for Twain himself was a bridge, a writer claimed by North and South, East and West, whose imagination spanned a terrain as vast as the United States itself. And I liked the idea of the Mark Twain Lighthouse, too, for what was Twain if not a beacon helping to give us our bearings?

The voice over the loudspeaker interrupted my reverie:

Two things you'd worry about going up and down the river. One was whether your boiler was going to overheat or not and the other was undesirables and pirates coming aboard ship to take your valuables. Well, we don't worry about a boiler, 'cause we don't have one on this boat. And I didn't *think* we had to worry about any undesirables, but it looks like the flood has left a couple over there in that shack. . . . We'll go over there and see what they're doing.

Sure enough, a couple of amateur actors dressed up as river bandits were making unconvincing threatening gestures on the shore. The whole improbable episode, which was spun out dully at great length, was a scheme to advertise that night's "Reflections of Mark Twain" at the Mark Twain Outdoor Theater. (Although the author who put sandwich boards on the knights of Camelot in *Connecticut Yankee* might have understood.) The boat edged closer to a group of islands near the Illinois shore. Finally the moment I'd been waiting for arrived. The guide announced over the loudspeaker,

The island coming up on our right is the most famous river island in the world, for it's the island Mark Twain wrote about. In *The Adventures of Tom Sawyer*, Mark Twain changed the names of several things. Hannibal became St. Petersburg, the widow Holliday became the Widow Douglas, his best friend Tom Blankenship became Huck Finn, and this island to our right became Jackson's Island. We are sure that this is Jackson's Island because if you put a couple of boys on a raft up around the highway bridge and sent them paddling for the Illinois shore, the current would set them down at just about the middle of this island. All these islands are owned by the federal government. You are free to camp and fish off them if you'd like. We have several varieties of wildlife on these islands—turtles, snakes, beaver, otter, muskrat. The Audubon Society tells us there are several varieties of birds. And at last count on Jackson's Island alone there were well over three billion mosquitoes.

End of comments.

I stared back at the Mark Twain Bridge as the boat returned to the landing. The tour guide, a genial local student, was standing near the exit as the passengers filed off.

"Do they tell you what to say," I asked him, "or can you say whatever you want?"

"I say whatever I want."

"Have you read *Huckleberry Finn*?"

"Sure. And *Tom Sawyer*, *Life on the Mississippi*—I read all that stuff to prepare for this job."

"Did you ever think of mentioning that Jackson's Island was also the scene of one of the most hopeful and positive black–white relationships in American literature?"

"Good point," he said. "Never thought of it. Maybe I'll put it in the tour next time. "

* * *

Passing through the *Mark Twain* riverboat's air-conditioned ticket office, I felt the sting of a wasp on my behind. My involuntary "ouch!" seemed to frighten a five-year-old boy with bangs who stood three feet behind me wearing what struck me as an oddly worried expression. He must have accidentally bumped into me, I remember thinking, although I couldn't for the life of me figure out how he could have caused that short, sharp sting.

"Did you hit her?" his mother demanded. The little boy started to bawl.

"Don't worry, It's okay, I'm fine," I reassured them. Then I saw what was in his hand.

The boy held a miniature version of the bullwhip that was the bestselling souvenir among schoolchildren at the Mark Twain Book and Gift Shop. If a child could create that sting with a miniature whip when he wasn't trying, I shuddered to think what a full-grown man could do with a full-sized whip when he *was* trying.

I turned to the mother of my five-year-old assailant. "Actually, it's *not* okay," I sputtered. "That's a bullwhip, and he should be taught not to use it on people!" She turned bright red, apologized, and snatched the whip out of the little boy's hand.

I had come to Hannibal to touch history, and history had eluded me at every turn. Until now. I had just been bullwhipped on the banks of the Mississippi.

All the anger that had been simmering in me for the four days I had been in Hannibal rose to the surface, but I didn't know whether to laugh or cry. I walked up Hill Street, heading back to my hotel. I passed the wax museum with the life-sized "Injun Joe" in the window and the souvenir shop that sold bullwhips to schoolchildren. Then I paused in front of the historic marker I had seen when I first arrived:

HERE STOOD THE BOARD
FENCE WHICH TOM SAWYER
PERSUADED HIS GANG TO
PAY HIM FOR THE PRIVILEGE
OF WHITEWASHING. TOM
SAT BY AND SAW THAT IT
WAS WELL DONE.

Tom Sawyer would be pleased as punch if he turned up in Hannibal today, I thought. He'd like that fierce-looking "Injun Joe" in the wax museum. He'd buy himself a bullwhip—crack it all over town, too. And he'd be right proud of how famous he was. If anyone asked him why he put Jim through all that misery at the end of *Huckleberry Finn*, he'd probably say, "Oh, pshaw, we were just having fun. Jim didn't mind." But we know he did. And we mind, too, particularly given the parallels between Tom's callous complicity in the re-enslavement of a free Jim and the machinations of whites in the 1880s, who were pushing free blacks in the South into a condition akin to slavery. What did Twain think of his famous hero? Well, when he called Theodore Roosevelt "the Tom Sawyer of the political world" he didn't mean it as a compliment. Roosevelt, like Tom, Twain wrote, was "always showing off, always hunting for a chance to show off . . . he would go to Halifax for half a chance to show off, and he would go to hell for a whole one."

Huck was willing to go to hell, too, but only if that was the price of freedom for his friend. Yes, this was Tom Sawyer's town all right. And the whitewashing was indeed well done.

Back at the hotel, I packed my clothes, books, and papers. I was about to toss out a *New York Times* from earlier in the week when I spotted a headline that read: RACIST YEARBOOK INCIDENT REVEALS RIFT IN GREENWICH. Five white high school seniors in Greenwich, Connecticut, had slipped a coded message into the school yearbook in which their initials in their photo captions spelled out "Kill all niggers." Does Greenwich have a racial problem? the reporter had asked residents. Yes, a black resident responded, but the town, she said, was "in total denial." Absolutely not, responded a white resident. "It's a lovely community, it really is. There is just everything here you could want. . . ." I thought about "the clammy stillness created and maintained by the lie of silent assertion—the silent assertion that there wasn't anything going on in which humane and intelligent people were interested." Another white resident interviewed by the *Times* felt that the guilty students—who were suspended—had been punished too harshly. "They're kids," she said. Like the kids wearing the "H.A.N.K." jackets at Hannibal High? Like the fun-loving kids who grew up to be the high-spirited law-enforcement agents who put up a sign at their annual convention in 1990 that read: NIGGER CHECK POINT. ANY NIGGERS IN THAT CAR?? 17¢ LB— and captured the hilarity of it all on film that somehow found its way into the national press in the summer of 1995? I checked out and loaded up the car. As I was leaving, the Twainland Express tootled by en route to "the best of Historic Hannibal."

*　*　*

A few hours before catching my plane back to Austin, I paid a call on Reverend Robert Tabscott, founder of the Elijah Lovejoy Society, in the St. Louis suburb of Webster Groves. Several people had mentioned his name during my various forays into Missouri black history, and I had had a stimulating conversation with him on the telephone. I decided to make his acquaintance before returning home.

Reverend Tabscott strode to the back room of the Elijah Lovejoy Society and emerged with a pen-and-ink sketch blown up to poster size and mounted on cardboard. "This is Meachum," he said. "It's the only picture I have of him. It came out of an old, old report to the public schools of St. Louis." He pulled out another sketch. "They had a contest in the 1940s to draw what you thought his 'Freedom School' looked like, and a little girl won five dollars for this picture of it." He read from a piece of paper he had removed from a file folder. "Now James Henry Turner says that Meachum 'was one of the wealthiest negroes of the city, and both he and my father were secretly interested in the Underground Railroad. . . .' "

A Presbyterian minister by trade, Reverend Tabscott, who is white, left his St. Louis pulpit in 1990 to research and teach the black legacy of St. Louis. Now he teaches courses on Missouri black history at Webster University, writes newspaper and radio columns, runs the Elijah Lovejoy Society (named for the martyred white abolitionist editor), and lectures in local primary and secondary schools. He also runs black-history field trips for local schools—"mostly to unmarked graves," he noted (adding that St. Louis interstate highways often run through black cemeteries). Black students' reactions to the tales he tells of St. Louis's black past range from "that sounds a lot like something my grandmother told me" to "why hasn't anyone told me about this before?" to "why am I denied my history?"

John Berry Meachum was a particular interest of his, and at one point he had envisioned having a statue of Meachum erected by the river. It hadn't happened. But that didn't surprise him. "Until the Dred Scott exhibit [at the courthouse] this summer," he said, "they denied there were ever slave sales on the steps of the courthouse."

During the two hours before I had to leave to catch my plane, I watched the tall, intense Presbyterian minister, who grew up in a southern Appalachian coal town in West Virginia, extract obscure books from the highest and lowest shelves of the packed bookcases along the length of the room, disappear into corners in which papers were piled three feet high, emerge with a yellowed article from the turn of the century, and dart to file cabinets for copies of radio and newspaper columns or videotapes he made for schools. The books, papers, news columns, and tapes all illuminated some aspect of the subject that this minister-without-portfolio (or congregation) pursued with indefatigable energy: black history in general and black history in Missouri in particular.

In a 1993 column, part of "Confessions of a White Man," a series on black history he did for the *St. Louis Post-Dispatch*, this fiftyish iconoclast wrote about why he does what he does. After the column appeared, someone called him and asked, "When are you going to decide what color you want to be?" A column he wrote in January 1994, "A Long History of Racial Intolerance," began with a description of an 1835 gathering of prominent St. Louis citizens that resolved to drive out of the state all "free Negroes . . . not born in Missouri," as well as "people suspected of holding views sympathetic to Negro causes." The column ended with a meditation on the lurid racist Ku Klux Klan graffiti that had been pasted onto the windows of the Elijah Lovejoy Society shortly before Reverend Tabscott received a telephone death threat. In the aftermath of this desecration of his window and the death threat (it was not the last), what disturbed him most was the following conversation he had had with a representative of a Webster Groves business group:

"People are concerned about what you are going to do next."
"Me?" [Tabscott] replied. "What have I done first?"
"Oh, you are on radio and you write for the newspaper."
"Is that a threat?" [Tabscott] asked.
Then the cryptic observation, "Well, a lot of colored come to your place."

"The ghouls in white sheets and hoods don't have to come again," he wrote in his column, since "their views are quietly held by so many people who live under the anonymity of respectability."

The only subject that seemed to vie with Missouri black history in Reverend Tabscott's pantheon of passions was Mark Twain. I ventured a question. "What do you think is American about Mark Twain?" Reverend Tabscott replied, "It's this extravagant, intoxicating ability to look at us and to depict the foibles, the romance, and the potential, all woven around the river which is always changing. I think the example of Twain for me that endures forever is when he's with Bixby," learning the piloting trade in *Life on the Mississippi*. He thinks he's learned the river, but then finds out he's also got to know it "in the night, in the dark . . . And Bixby says, 'You've got to learn it all over again.' And that is to be an American—if we are to survive. You've got to learn it all over again—but it's got to be learned with a new rubric of a deeper fascination for a history that is so conflicted and so wonderful and so tragic and so sublime."

You learn it all over again. I like that, I thought. Yes, that's an excellent thing to do with a history as filled with unexpected depths and hidden reefs and shadows and shoals as Mark Twain's river—and as Mark Twain himself. You learn it all over again. . . . Changes in circumstance and awareness can make everything we thought we understood look different. What else is there to do besides go back and start all over? Wasn't that what Twain himself had done when he returned to the scenes of his childhood as someone who asked questions that child had never asked?

2

EXCAVATIONS

August 1995. It was the kind of event that ordinarily would not have made it into the newspapers at all. But Johnson C. Whittaker's commission as a second lieutenant in the U.S. Army was coming a bit late—115 years late. And press accounts of the ceremony at the White House showed the federal government facing the same challenge the Southern Baptist Convention had struggled with two months earlier: apologizing for past racism. As I read the reports, it dawned on me that I had first heard of Whittaker's case years earlier, through Mark Twain's comments on it—although he hadn't mentioned Whittaker by name. This was a denouement Twain would certainly have followed with interest.

In 1880 Johnson Whittaker was summarily expelled from West Point. The press largely accepted the official line that he had been guilty of "conduct unbecoming an officer." Twain had no access to evidence exonerating the cadet, and it is not clear how much he knew about the details of the case. What angered him, though, was the assumption—which pervaded the press accounts—that it was because Whittaker was "black" that he had acted irresponsibly, a corollary of the racist belief that members of an "inferior" race were not the stuff of which officers were made. William Dean Howells recalled Twain's response:

> About that time a colored cadet was expelled from West Point for some point of conduct "unbecoming an officer and gentleman," and there was the usual shabby philosophy in a portion of the press to the effect that a negro could never feel the claim of honor. The man was fifteen parts white, but, "Oh yes," Clemens said, with bitter irony, "it was that one part black that undid him." It made him a "nigger" and incapable of being a gentleman. It was to blame for the whole thing. The fifteen parts white were guiltless.

Some ten years later Twain again showed his skepticism about this racist line of argument in a passage in the manuscript of *Pudd'nhead Wilson*: Tom Driscoll, who has just come to understand that he is one thirty-second part black, recognizes that "that which was base" in his character came not from his black blood but rather from his white blood—"the white blood in him debased by the brutalizing effects of a long-drawn heredity of slave-owning, with the habit of abuse which the possession of irresponsible power always creates & perpetuates."

Born a slave in South Carolina in 1858, Johnson Whittaker was appointed to West Point in 1876. After his freshman-year roommate Henry O. Flipper left in the spring of 1877—the first black to graduate from West Point—Whittaker was the only black cadet. Isolated, ostracized, and cut off from student life, Whittaker took solace in daily Bible readings. When, during his second year, he wrote in his Bible, "Try never to injure another by word, by act, or by look even. Forgive as soon as you are injured, and forget as soon as you forgive," he probably never anticipated having to forgive anything like the brutal racist assault he would suffer two years later. In the predawn hours of April 6, 1880, three masked intruders attacked him in his sleep, beating him, slashing his ears ("like we do hogs down South," one cadet remarked), smashing a mirror on his head, tying his arms and legs to his bed, and threatening to kill him if he called for help. They left him bleeding in the dark in his underwear. He was found unconscious later that morning. No white cadets confessed to the crime. West Point officials charged Whittaker with having staged the attack and mutilated himself to discredit the military and win sympathy, since he expected to fail an exam and be expelled. In the court of inquiry and court-martial, the prosecutor referred to the "inferior" and "superior" races and asserted that "Negroes are noted for their ability to sham and feign." The *New York Times* quoted cadets as saying, "Oh, it's just like a nigger, you know," and "Niggers are capable of anything." Later, on the same day that President Chester Arthur ruled that the court-martial had been illegal, Secretary of War Robert T. Lincoln, son of the president whose Emancipation Proclamation had freed Whittaker, discharged him anyway, allegedly for having failed a philosophy exam.

Whittaker went on to become a lawyer, teacher, and school administrator. He rarely talked about his West Point experience; according to his granddaughter, he "burned all the papers about it because he did not want his sons to grow up bitter." Both of his sons served as commissioned officers in the army during World War I, and during World War II his grandson was a fighter pilot with the celebrated Tuskegee Airmen.

In the summer of 1995 Johnson Whittaker finally got his due. President Clinton called him "a man who through courage, example and perseverance, paved the way for future generations of African-American military leaders." Senator Ernest "Fritz" Hollings of South Carolina, who had set in motion the investigation that culminated in the White House ceremony, welcomed the chance to "rectify a grievous injustice." Even Senator Strom Thurmond of South Carolina rallied to Whittaker's cause, saying that the commission sent "a message that injustice and discrimination will not be tolerated within a democracy." But as Mark Twain well knew, injustice and discrimination had been part and parcel of our democracy from the start. Indeed, some have argued that that awareness is at the center of what makes Twain's greatest work so valuable. A student of Johnson Whittaker's when Whittaker was a principal in Oklahoma City went on to become a leading proponent of this view. Mark Twain, he maintained, grasped "the moral situation of the United States and the contrast between our ideals and our activities. . . . One of the functions of comedy is to allow us to deal with the unspeakable. And this Twain did consistently."The "unspeakable" that Twain helped us address was slavery and its legacy—a combustible and unavoidable topic even today. The student, who became one of the most careful and sensitive readers of Twain in our time, was Ralph Ellison.

How did Twain come to understand the unspeakable betrayals at the heart of American history? How did he come to construct great art out of them? The letters seventeen-year-old Sam Clemens sent home when he first left Hannibal for the East in 1853 show a young man who expressed attitudes then typical of the majority of people in his hometown: he was rabidly antiabolitionist and he assumed white superiority to be part of the natural order of things. He wrote his mother about a courthouse "surrounded with chains and companies of soldiers to prevent the rescue of McReynold's niggers, by the infernal abolitionists," and he groused that "in these Eastern States niggers are [treated] considerably better than white people." Within sixteen years, however, he would write an editorial blasting assumptions of white superiority—a theme echoed in his comments about Johnson Whittaker. Within twenty-one years he would publish a story that ridiculed white myopia about black experience. And within twenty-three years he would begin a novel that ranks as one of the most scathing critiques of racism by an American. How did he get from there to here?

The road from Hannibal to Hartford, I knew, led through Elmira, New York, a place I had never visited. For this reason I was pleased to accept an invitation from the Center for Mark Twain Studies to lecture there in September 1995. The physical distance I traveled in a day—flying from Austin to Dallas to New York to Syracuse to Elmira—was nothing compared to the moral and social distance Mark Twain had covered between Hannibal and *Huck Finn*. I hoped Elmira would shed some light on this extraordinary journey.

*   *   *

September 10, 1995. As my plane made its way through the pitch-black skies from Syracuse to Elmira, I thought about what forces had converged to bring Mark Twain to this part of the world more than a century ago. Among them, certainly, was his taking a berth on the *Quaker City* and meeting up with a boon traveling companion who carried a striking cameo of his beautiful sister. Who knows, but were it not for this minor detail of the grand design, the man who hailed from Hannibal and points west might never have landed in the Langdon parlor at all.

Hannibal was probably far from Sam Clemens' mind in 1868 as he applied to Mr. and Mrs. Jervis Langdon of Elmira for permission to marry their daughter Olivia. He had left Hannibal some fifteen years earlier, at age seventeen, returning only briefly in the interim. Since his departure, he had traveled widely in the United States and abroad. The Western sketches he had published while living in San Francisco had won him a rising reputation as a comic writer. His newspaper reports about the Grand Tour he had just completed of Europe and the Holy Land further enhanced his reputation, and he was currently in the process of reshaping that material into a lavishly illustrated subscription-market book. Olivia's brother, Charles, who had been a fellow passenger on the *Quaker City* tour, had introduced Clemens to the family and vouched for his friend's worthiness. The Langdons, however, had heard rumors of Clemens' free-wheeling lifestyle in the West. Was he an irresponsible drunk? Was he of sufficiently firm moral fiber? They asked for references. Clemens supplied names and addresses. The Langdons sent out letters. The replies came back mixed—a wholly unsatisfactory grab bag of equivocation, insult, and faint praise. "Haven't you a friend in the world?" Jervis Langdon asked incredulously. "I'll be your friend myself," he said with a generosity that would continue to give Twain ample reason to marvel. This brash young man so deeply in love with his daughter, who hadn't the sense or sophistication to command a decent batch of letters, had won the heart of the father as well and was welcomed into the family.

Jervis Langdon was the polar opposite of Mark Twain's own father. John Marshall Clemens was considered cold and aloof, Langdon was open and warm. John Marshall Clemens was a slaveholder, Langdon an abolitionist. Every business venture John Marshall Clemens touched turned to dust; everything Langdon touched throve and prospered. Jervis Langdon did more than allow Twain to marry his daughter. He bought him an elegant, fully furnished house and helped him buy a part interest in the *Buffalo Express*; he also dispensed advice with sagacity and affection. Twain came to refer to him simply as "Father." During the brief period between Twain's first trip to Elmira in 1868 and his father-in-law's death in 1870, Langdon became just that: a new father, a beloved surrogate father, perhaps the father Twain wished he had had. Both morally and materially, Jervis Langdon was central to the process by which Sam Clemens remade himself into Mark Twain. The heritage of proud antislavery activism Langdon embodied helped goad Twain to raise the questions he had never asked as a child. And the sprawling farmhouse Langdon bequeathed to his daughter Susan Langdon Crane was an ideal place to ask them. It would be home to Twain during his most productive periods as a writer. It was in Elmira, nine hundred miles from the scenes of his childhood, that Mark Twain recalled and retrieved the Matter of Hannibal as a subject for his art.

* * *

My plane landed. Twain scholar Michael Kiskis met me at the airport and drove me over a series of winding, wooded roads that led to Quarry Farm, where Twain had spent almost every summer from 1871 to 1889. Through the generosity of Jervis Langdon Jr., who ten years ago gave the property to the Center for Mark Twain Studies at Elmira College, visiting scholars can now stay at Quarry Farm. I was greeted by the

6. *Jervis Langdon, Mark Twain's abolitionist father-in-law, who both morally and materially was central to the process by which Sam Clemens remade himself into Mark Twain (Photo courtesy of the Mark Twain House, Hartford, Connecticut)*

caretaker, who presented me with a welcome folder, a key to the farm-house, and instructions on how to reach her in an emergency. Then she left me on my own.

The lights of Elmira sparkled in the distance as I swayed in a rocking chair on the porch where Mark Twain used to sit. It was here, at Quarry Farm, that Twain wrote *Adventures of Huckleberry Finn* and other major works. The night air grew chill. I went inside.

"Explore the house, enjoy yourself," said Gretchen Sharlow over the phone shortly after I arrived. She was the director of the center and my gracious host. The leaflet she left me said there were twenty-five rooms. I had them all to myself. I walked through the sixteen I could find, improvising a silent sound track at every step: "This is the table where they ate breakfast. . . . Maybe Twain sat in this chair. . . . This is the room Twain slept in, dreamed in." I giggled to myself as I recalled the fun Twain had had in *The Innocents Abroad* with the guide in Genoa who aspired (without success) to elicit from the American tourists just this kind of excitement about what Christopher Columbus had touched.

7. *Quarry Farm, the house in Elmira, New York, where Mark Twain and his family spent summers during his most productive years as a writer (Photo courtesy of the Mark Twain House, Hartford, Connecticut)*

I knew I would have to rise early: I had research to do at the Chemung County Historical Society, the public library, and the college library; I wanted to track down and interview a distant relation of Mary Ann Cord's; and I needed some time to go over my public lecture scheduled for the following night. But I also knew I was not going to bed any time soon—not with all those books within my reach! Summers at Quarry Farm were times when Twain did a lot of reading—alone, with family members, aloud to the family circle and to groups of friends and neighbors. Many of the books that were in the house when he lived there were still sitting on the shelves. Those known to have markings by Twain had been removed to the college library's Twain collection, but hundreds of others remained. "Feel free to read the books," the caretaker had told me. I felt like the proverbial kid in a candy store.

I glanced at some of the elaborate sets: *Lives of the Queens of England from the Norman Conquest with Anecdotes of their Courts*, by Agnes Strickland; *Lives of the Queens of England Before the Norman Conquest*, by Matthew Hall; *Charles Knight's Pictorial History of England, From 56 B.C. to A.D. 1867*. Maybe Mark Twain had consulted these to create his history game, an aerobic history lesson he designed for his children, where players had to shout out the reigning kings and queens as they ran between markers designating years on the long driveway.

In the comfortable, elegantly furnished parlor and the room next to it were sets of Shakespeare, Browning, Cowper, Goethe, Pope, George Eliot, Macaulay, and Carlyle—two sets of Carlyle, to be precise. (He seems to have been a family favorite of the Langdons. Olivia had been reading him since shortly before she turned eighteen in 1863, when she copied the following into her commonplace book: "Next to possessing genius one's self is the power of appreciating it in others.—Carlyle.") One set, lavishly published in Boston by Estes and Lauriat in 1885, was large and ornate, with a heavy, gold-embossed black leather binding and marbled endpapers. The other set, a much cheaper one published by Scribner, Welford in New York in 1871, was pocket-sized and bound in red cloth. I took one of the trim little volumes off the shelf—*The French Revolution*, volume 1, a book Twain had called "one of the greatest creations that ever flowed from a pen" and claimed to have read eight times; it was also the book he chose to read minutes before he died. Twain said he had read the book for the first time in 1871, the year of publication of the volume I held in my hand. It was the size of the paperback mystery I finished on the plane. Could this be the copy he read? Could Twain have stashed this book in his pocket and taken it outside on the porch, or to his hilltop study, or to the portable hammock on the lawn, where he would read and discuss books on summer afternoons with his brother-in-law? I carefully turned the pages. I froze when I got to page 160. A passage was clearly marked in pencil in the left-hand margin, a phrase was enclosed within penciled brackets, and another was underlined boldly with an even, sharp stroke. A shiver of excitement went through me as I read.

O poor mortals, ⟨ how ye make this Earth bitter for each other; ⟩ this fearful and wonderful Life fearful and horrible; and Satan has his place in all hearts! Such agonies and ragings and wilings ye have, and have had, in all times:—to be buried all, in so deep silence; and the salt sea is not sown with your tears.

Great meanwhile is the moment, when tidings of Freedom reach us; when the long-enthralled soul, from amid its chains and squalid stagnancy arises, were it still only in blindness and bewilderment, and swears by Him that made it, that it will be *free!* Free? Understand that well, <u>it is the deep commandment, dimmer or clearer, of our whole being to be *free*.</u> Freedom is the one purport, wisely aimed at, or unwisely, of all man's struggles, toilings and sufferings, in this Earth. Yes, supreme is such a moment (if thou have known it): first vision as a flame-girt Sinai, in this our waste Pilgrimage,—which thenceforth wants not its pillar of cloud by day, and pillar of fire by night! Something it is even,—nay, something considerable, when the chains have grown corrosive, poisonous,—to be free "from oppression by our fellow-man." Forward, ye maddened sons of France; be it toward this destiny or toward that! Around you is but starvation, falsehood, corruption and the claim of death. Where ye are is no abiding.

Intuitively I knew Twain had been here. The next day I notified Mark Woodhouse, the college archivist, who came to take the volume to the library. He confirmed my suspicions: the markings were identical to scores of others verified by experts as having been made by Twain. Why did he mark this passage and emphatically underline "<u>it is the deep commandment, dimmer or clearer, of our whole being to be *free*</u>"? Before 1871 he had written little that might even remotely be connected with the theme of freedom; but that theme would soon come to animate his greatest work of fiction. Might there have been something in his immediate surroundings in Elmira that helped open his eyes to its significance?

Jervis Langdon had been a "conductor" on the Underground Railroad, and Elmira had been a key "stop," the main point of transfer, in fact, between Philadelphia and Saint Catharines, Ontario. Escaped slaves first began to come to Elmira via the Underground Railroad about 1840, mainly from Maryland and Virginia. Sometimes they followed the Susquehanna River, "up from Chesapeake Bay to Williamsport and then traveling across low hills and down the valley to South Creek into Elmira," writes Chemung County historian Winifred Eaton. "For these fugitives Elmira was a 'gateway to freedom': some used it as a way station on their route to Canada; others stayed here, feeling that the town was a friendly refuge." Slaves who made it as far as Philadelphia would often be given sufficient funds to reach Elmira by Philadelphia Anti-Slavery Society leader William Still. Once they arrived, usually in parties of six or seven, they would seek out John W. Jones, an ex-slave who had escaped from Virginia in 1844 and was the leading coordinator of Underground Railroad activities in Elmira. Often the slaves were "entirely penniless . . . and money had to be obtained to send them on their way." Jervis Langdon, who had made a fortune in the coal industry, was Jones's chief financial supporter. His involvement was noted by Wilbur Siebert, whose 1897 history of the Underground Railroad mentions Langdon, along with Jones, as a Chemung County operator. Jones would get the escaped slaves to a train that left Elmira at three or four in the morning. Sympathetic railway employees would ensconce the fugitives in the baggage car and transport them safely to Canada. Eight hundred fugitive slaves were cared for in this way, not one of whom, to Jones's best knowledge, was ever recaptured.

Although Langdon's money was particularly important to the smooth operation of the Underground Railroad in Elmira, his role was more than merely financial. Langdon, together with Hiram Crane (father of Theodore, Langdon's business manager and future son-in-law) and Dr. Nathan Smith and his wife, would hide slaves on farms in the countryside surrounding Elmira. Once a local judge secretly called Langdon into his chambers and asked him to warn a group of fugitive slaves hiding near town that a slave-catcher who had been pursuing them was about to reach Elmira. While living in the upstate New York town of Millport in 1838, Langdon helped Frederick Douglass escape from slavery.

Prominent abolitionists, including William Lloyd Garrison and Gerrit Smith, had been guests at the Langdons', and there is evidence that Langdon may have supported Underground Railroad activities in the vicinity of Cairo, Illinois. Friends of the Langdons in Elmira had opened their homes to fugitive slaves. Indeed, John Jones became a member of the household of Judge Ariel Thurston, whose sister "conducted a private school, where [Jones] was permitted to be a pupil." The woman who taught John Jones to read and write would later supervise Susan and Olivia Langdon's studies at an establishment that the Langdon girls affectionately called "Miss Thurston's." After the war Langdon was active in behalf of the freedmen, embracing such causes as the education of blacks in the South.

Although slavery had been abolished three years before Twain arrived in Elmira, reminders of the city's role as a center of antislavery activity were still plentiful in 1868. For example, in 1846—the same year the Southern Baptists broke away to create a haven for slaveholders—Langdon and forty other Elmirans founded the Independent Presbyterian Church (renamed Park Church in 1874) when the First Presbyterian Church refused to take a strong stand against slavery. They adopted a number of bylaws for the new church, including this one:

> That the using, holding, or trading in men as slaves is a sin in the sight of God, a great wrong to its subjects and a great moral and political evil inconsistent with the Christian profession. And that this church will admit no person into its pulpit or communion who is known to be guilty of same.

Twain must have been told something of this history when the Langdons first took him to their church. Whether or not Jervis Langdon talked with Twain about these times and these events himself, his abolitionist activities were prominently recalled by many, including Douglass, at his death in 1870; it is inconceivable that Twain was not deeply aware of them. When Langdon died, Douglass wrote his widow, Olivia Lewis Langdon, "If I had never seen nor heard of Mr. Langdon since the days that you and himself made me welcome under your roof in Millport, I should never have forgotten either of you. Those were times of inefface[a]ble memories with me, and I have carried the name of Jervis Langdon with me ever since." (Langdon most likely also played a central role in the establishment of one of the main black churches in Elmira, built around the time of the Civil War with the help of "generous white friends" and later known as the Douglass A.M.E. Zion Church.)

Twain must have been struck by the distance between his real father and his chosen surrogate father on the issue of slavery and abolition. While one was sending abolitionists to the state penitentiary, the other was helping to fund their activities. Louis J. Budd suggests that it may have been, in part, to win his in-laws' esteem that Twain penned an antilynching editorial in the *Buffalo Express* in 1869. Twain's first biographer argued for a more fundamental transformation on Twain's part.

Although slavery had been outlawed, racism was still perfectly legal. Racially motivated lynchings, police brutality, and denials of justice were alive and well and would soon come to play a central role in Twain's work. As Paine notes, Twain's contributions to the *Buffalo Express* and *Galaxy Magazine* in 1869 and 1870 were less likely to be humorous sketches and more likely to be "either savage assaults upon some human abuse, or fierce espousals of the weak. They were fearless, scathing, terrific." In two pieces—"Only a Nigger" (dealing with a recent lynching in Memphis) and "Disgraceful Persecution of a Boy" (concerning the treatment of the Chinese in San Francisco)—Twain satirically explored the subject of unexamined racism, deploying the rhetoric of the bigot in order to expose him. In "Only a Nigger," for example, Twain wrote,

> Ah, well! Too bad, to be sure! A little blunder in the administration of justice by Southern mob-law; but nothing to speak of. Only "a nigger" killed by mistake—that is all. . . . But mistakes will happen, even in the conduct of the best regulated and most high-toned mobs, and surely there is no good reason why Southern gentlemen should worry themselves with useless regrets, so long as only an innocent "nigger" is hanged, or roasted or knouted to death, now and then. What if the blunder of lynching the wrong man does happen once in four or five cases? Is that any fair argument against the cultivation and indulgence of those fine chivalric passions and that noble Southern spirit which will not brook the slow and cold formalities of regular law when outraged white womanhood appeals for vengeance? Perish the thought so unworthy of a Southern soul! Leave it to the sentimentalism and humanitarianism of a cold-blooded Yankee civilization! What are the lives of a few "niggers" in comparison with the impetuous instincts of a proud and fiery race? Keep ready the halter, therefore, o chivalry of Memphis! Keep the lash knotted; keep the brand and the faggots in waiting, for prompt work with the next "nigger" who may be suspected of any damnable crime! Wreak a swift vengeance upon him, for the satisfaction of the noble impulses that animate knightly hearts, and then leave time and accident to discover, if they will, whether he was guilty or no.

Twain knew that the South had no monopoly on racism, as is apparent from a piece he published in the *Buffalo Express* the following year, in which he condemned a Buffalo police officer's use of unwarranted force against a "slightly inebriated and noisy" black man who died as a result. (The incident Twain described with pointed sarcasm was virtually identical to an incident that took place in Mississippi eighty years later, which engendered Anthony Walton's powerful 1996 book *Mississippi: An American Journey.*)

In "Disgraceful Persecution of a Boy" Twain writes from the standpoint of an outraged editorialist: "In San Francisco, the other day, 'a well-dressed boy, on his way to Sunday school, was arrested and thrown into the city prison for stoning Chinamen.' . . . What had the child's education been," he asks, and proceeds to argue that the child should not be blamed for doing exactly what his elders taught him to do.

> It was in this way that the boy found out that a Chinaman had no rights that any man was bound to respect; that he had no sorrows that any man was bound to pity . . . everybody, individuals, communities, the majesty of the State itself, joined in hating, abusing, and persecuting these humble strangers. And, therefore, what *could* have been more natural than for this sunny-hearted boy, tripping along to Sunday school, with his mind teeming with freshly-learned incentives to high and virtuous action, to say to himself:
>
> "Ah, there goes a Chinaman! God will not love me if I do not stone him."
>
> And for this he was arrested and put in the city jail. Everything conspired to teach him that it was a high and holy thing to stone a Chinaman, and yet he no sooner attempts to do his duty than he is punished for it.

Both here and in "Only a Nigger" it is people who think of themselves as "good"—a little boy on his way to Sunday school; fine upstanding Southern gentlemen—who are the perpetrators of evil. "Defining right behavior and impeaching bad had of course been a part of his satiric agenda from the beginning," writes Jeffrey Steinbrink, but in the writing Twain did during the spring of 1870 for the *Galaxy* and the *Buffalo Express* "social criticism emerged unambiguously as a focus." Whether he wrote to please his in-laws (as Budd suggests), to craft his professional image (as Steinbrink argues), or because he now genuinely embraced these views himself (as Paine believes), there was a marked change in the kinds of issues Twain chose to address and the manner in which he addressed them. But one subject that he still had not explored in print was the Matter of Hannibal and the matrix of personal and moral complexities it entailed.

The 179 households that made up Elmira's black community when Twain arrived there in 1868 included many former slaves, each with his or her own compelling story. Twain's active interest in such stories is evidenced by the care with which he recorded one of them in black ink on the front flyleaf of his copy of William Still's book, *The Underground Rail Road: A Record of Facts, Authentic Narratives, Letters, &c., Narrating the Hardships, Hair-breadth Escapes, and Death Struggles of the Slaves in Their Efforts for Freedom, by Themselves and Others, or Witnessed by the Author.*

Mrs. Luckett was a slave in Richmond, with a daughter 3 years old. Her brother, Jones, an escaped slave, lived in Elmira (1844). [This may well have been John W. Jones, Jervis Langdon's colleague on the Underground Railroad.] He cut two duplicate hearts out of pink paper, & wrote on one, "When you see this again, you will know." No other word accompanied it. After a while a white man went [to] Richmond with the other heart, called on the woman's mistress on some pretext which brought in the slaves. "Mrs. L. saw & recognized the duplicate heart; she escaped, with her child in the night, joined the man at a place appointed, (Annapolis,) & thence got through safely to Elmira. She lives in Canada, now (whither she had to flee when the fugitive slave law was passed [1850,]) & the child is also married & lives in Binghamton, N.Y., (1884.) This account given by Mother [Olivia Lewis Langdon, Clemens' mother-in-law], who knew the several parties.

Each story was different—but each testified to "the deep commandment, dimmer or clearer, of our whole being to be *free*."

* * *

September 12, 1995. Donald Blandford carefully unfolded the torn, yellowing newspaper article, dated Sunday, April 3, 1938, and spread it out on the vinyl tablecloth in Green Pastures, a bar and restaurant owned by Elmira jazz musician Howard Coleman. ELMIRAN, 80, FORMER SLAVE read the beginning of the headline (the remainder of which had been left behind when the article was ripped from the *Sunday Telegraph Elmira Star-Gazette* years ago). The stately octogenarian in the picture that accompanied the story was Blandford's grandfather, Alsace Blandford, an artist who had been born a slave in Maryland.

Alsace's father, a slave named Thomas Blandford, had had unusual freedom of movement as drayman on a plantation in Prince George's County, Maryland, where one of his regular tasks was delivering his master's crop to market in Baltimore. One day he didn't stop at Baltimore but "kept right on going." He had built a crawl space under a layer of timbers in his wagon, and he smuggled his wife and four children out of slavery under a load of potatoes. One of the children under those potatoes was Alsace, who moved to Elmira in 1880, when he was twenty-two, and took up residence near Quarry Farm in a house at 811 East Avenue that Mark Twain passed on his daily walk. Alsace spent the rest of his life in Elmira, where his extended family would come to include another ex-slave named Henry Crummell Washington.

Henry Washington followed the North Star to freedom in 1858, when he was about thirteen years old, and ended up in Elmira. His grandson recalls his having told the story of "sitting by the Erie Railroad platform when the man who ran the barbershop in the Rathbun Hotel befriended him, gave him a home, and taught him how to barber." Ten years later, when Mark Twain arrived in Elmira, city records indicated that many masons, teamsters, grocers, shoemakers, and virtually all of the barbers in town were "colored." Henry Washington became one of the best known barbers in Elmira. Members of Elmira's elite—including Jervis and Charles Langdon and Twain—were among his customers. They even had their own shaving mugs on a rack in his shop. It is likely that Twain had his hair cut by Henry Washington before he knew the story of his life, a story that would have enormous impact on Twain's career as a writer.

One evening in the summer of 1874, as the sun set over the Chemung River valley, a former slave named Mary Ann Cord, the cook at Quarry Farm, told Twain, Livy, and others assembled on the porch the powerful story of how she was forcibly separated from her husband and children on the auction block and eventually reunited with her youngest child, Henry Washington, after the war. As Twain would recall in an unpub-lished memoir, "she had the best gift of strong & simple speech that I have known in any woman except my mother. She told me a striking tale out of her personal experience, once, & I will copy it here—& not in my words but her own. I wrote them down before they were cold." Mary Ann Cord's story stunned and moved Twain. He found it "a shameful tale of wrong & hardship," but also "a curiously strong piece of literary work to come unpremeditated from lips untrained in literary art." It be-came the basis for the first piece Twain published in *The Atlantic Monthly*, "A True Story, Repeated Word for Word as I Heard It."

Aunt Rachel, the name Twain gives Cord in the piece, tells her story with great power and directness.

8. *Mary Ann Cord, a former slave whose storytelling stimulated Twain to revisit scenes of slavery as a writer (Photo courtesy of the University of Maryland, College Park)*

Dey put chains on us an' put us on a stan' as high as dis po'ch,— twenty foot high,—an' all de people stood aroun', crowds an' crowds. An' dey'd come up dah an' look at us all roun', an' squeeze our arm, an' make us git up an' walk, an' den say, "Dis one too ole," or "Dis one lame," or "Dis one don't 'mount to much." An' dey sole my ole man, an' took him away, an' dey begin to sell my chil'en an' take *dem* away, an' I begin to cry; an' de man say, "Shet up yo' dam blubberin'," an' hit me on de mouf wid his han'. An' when de las' one was gone but my little Henry, I grab' *him* clost up to my breas' so, an' I ris up an' says, "You shan't take him away," I says; "I'll kill de man dat teches him!" I says. But my little Henry whisper an' say, "I gwyne to run away, an' den I work an' buy yo' freedom." Oh, bless de chile, he always so good! But dey got him—dey got him, de men did; but I took and tear de clo'es mos' off of 'em an' beat 'em over de head wid my chain; an' *dey* give it to *me*, too, but I didn't mine dat.

"A True Story" confounded readers and commentators. Quite simply, they didn't know what to make of it: they expected humor from Twain, and looked for a joke lurking somewhere. But there was no joke. Instead, there was black dialect, presented not to be condescended to or ridiculed, as it so often was in typical dialect tales of the time, but to be received respectfully as a vehicle for communicating raw pain, strong emotion, harsh truth. Cord's story condensed the pain of the thousands of other ex-slaves who tried—and often failed—to reunite their families. This woman, who had been "in slavery more than forty years," as Twain knew, brought back to him, in her image of children being wrenched from their mother's arms on the auction block, the agony inflicted by slavery and the enigma of human cruelty. As she jolted him into recollections of scenes he had long banished from his mind, Mary Ann Cord helped put the Matter of Hannibal on Mark Twain's porch in Elmira.

Twain would revisit the auction block in first person in chapter 34 of *A Connecticut Yankee*, which is entitled "The Yankee and the King Sold as Slaves." The "auction block came into my personal experience," Hank Morgan tells us, and "a thing which had been merely improper before became suddenly hellish." And Mary Ann Cord's experiences would also be echoed in chapter 21 of *A Connecticut Yankee*, in Twain's unflinching description of a slave mother being torn from her child and then brutally lashed: "She dropped on her knees and put up her hands and began to beg, and cry, and implore, in a passion of terror, but the master gave no attention. He snatched the child from her . . . then he laid on with his lash like a madman till her back was flayed, she shrieking and struggling the while piteously."

Mary Ann Cord's grief and despair at being separated from her family, and the language in which she expressed those emotions, would find their way into chapter 23 of *Huckleberry Finn*, where Huck describes Jim "setting there with his head down betwixt his knees, moaning and mourning to himself." Huck tells us,

> I knowed what it was about. He was thinking about his wife and his children, away up yonder, and he was low and homesick; because he hadn't ever been away from home before in his life; and I do believe he cared just as much for his people as white folks does for theirn. It don't seem natural, but I reckon it's so. He was often moaning and mourning, that way, nights, when he judged I was asleep, and saying, "Po' little 'Lizabeth! po' little Johnny! it mighty hard; I spec' I ain't ever gwyne to see you no mo', no mo.' "

Jim's pain, like Mary Ann Cord's, is palpable. Their feelings run directly counter to prevailing wisdom as enshrined in the *American Cyclopaedia*, a popular multivolume reference work which Twain (at least on other points) consulted often. Under the entry "Negroes" the encyclopedia stated, "Negroes . . . are comparatively insensible to pain." (Or, as Thomas Jefferson had put it, "Their griefs are transient.")

Perhaps the most important thing Mary Ann Cord taught Twain was that the vernacular—in her case the ungrammatical dialect of a woman with no formal education—could move a narrative forward dramatically and effectively. While other speakers of his acquaintance had helped make Twain aware of the vitality of the spoken word and of the challenge of transferring it to print, Mary Ann Cord helped spark his awareness of the potential of a vernacular voice as the scaffolding of powerful fiction.

* * *

In 1881 Frederick Douglass published his *Life and Times*, the third of his autobiographical works. As noted by his most recent biographer, William McFeeley, hidden amidst the retellings of familiar stories from his life and observations on the Civil War and its aftermath was "the book's real message—which few people received," namely, "that the story of slavery should not be purged from the nation's memory. White America wanted to hear no more of the subject; emancipation had taken care of it. Many black Americans, reacting to this weariness, had become almost apologetic about their slave past." McFeeley writes, "The book [was] a cry from the heart." It sold few copies. But the idea that slavery and its tragic consequences "should not be purged from the nation's memory" found a strong and perhaps unexpected ally in the person of Mark Twain, who at this point had written mainly comic sketches, travel books, and books for children. *Adventures of Huckleberry Finn*, published in 1885, would manage to keep American slavery on the table in classrooms and libraries on every continent for the next hundred years and more. And the turn that novel took as Twain finished it in the early 1880s ensured that it would also keep reminding Americans of the travesty that freedom had become by the late 1870s for so many of the former slaves.

Stories like Mary Ann Cord's were central to the genesis of *Huckleberry Finn*. I had wondered whether any of her descendants were left in town, and I was delighted when I found Donald Blandford, the oldest living member of Cord's family still in Elmira, who agreed to get together with me at Green Pastures. Donald Blandford had quite a few stories of his own to tell, and Gretchen Sharlow and I listened to his reminiscences for much of the afternoon. When he returned to Elmira after serving in the army, he went to work at the Mark Twain Hotel. The best job he could get was busboy, cleaning glasses and setting up the soda bar. "People would come in and ask me to make them a banana split or a milkshake or something and I couldn't do it." Mary Ann Cord had prepared nearly all the food and drink that Mark Twain consumed while in Elmira, but Blandford wasn't allowed to touch the soda pump. Only whites could be soda jerks. "I had wanted to become a journalist," he reflected, "but got sidetracked because of what they call racism today." Although he was better qualified than many of the white reporters already on the staff of the local paper, the editors insisted that he did not meet their hiring standards. (I was reminded of a story I'd heard the day before from Robert and Helen Jerome. In the summer of 1958 Robert Jerome had a minimum-wage job open at the family retail clothing store, but he was embarrassed even to suggest it to a personable young man whom he knew and liked as a customer, a talented black high school senior and a star athlete. After all, the young man had been recruited heavily by many colleges but had chosen Syracuse University, and there were many Syracuse alumni among Elmira businessmen. Surely he would be swamped with better offers. No, the young man told him. He had tried. All the local businesses had turned him down. Robert Jerome asked around and found that " 'We can't' was always the excuse." Prejudice, Jerome believes, was the only reason. He gave the young man a job and worked alongside him in the store that summer. Shortly thereafter Ernie Davis became the first black to win the Heisman Trophy.)

Donald Blandford eventually took a job at the local electric plant. He became active in the electrical workers' union and in community affairs, serving as president of the Elmira branch of the NAACP, and channeled his interest in journalism into a newsletter. What kinds of things did he write? I asked. He quoted a few lines from one of his columns from the 1960s: "Blacks are tired, angry. Tired of taking the white man's leavings. . . ." He quickly got a reputation for being "inflammatory," he said, smiling at the memory. What writers had inspired him? What writers did he admire? James Baldwin, he said. Richard Wright. He was also a passionate fan of Mark Twain. *Pudd'nhead Wilson* was his favorite. He liked the way Twain used fingerprinting and the ploy of the switched babies to expose racism.

Donald Blandford was proud of his family connection to Mary Ann Cord and had brought along a folder full of photocopies about her: "A True Story" copied from the first book in which it was published, *Sketches, New and Old*; articles about the inscription Twain wrote on the flyleaf of the book he presented to Cord and about the book's being deposited at the University of Maryland library in 1986. "You might be interested in these," he said, pulling out some photographs. "I took them earlier today with my grandson." They were photographs of Mary Ann Cord's grave in the Elmira cemetery, as well as the graves of other relatives. A large headstone bearing the family name WASHINGTON adorned the plots, with the birth and death dates for Mary Ann Cord; her son, Henry Washington; her granddaughter, Louise Washington Condol; and other family members. Donald Blandford had been meaning to go there for a long time, he said, but hadn't until today, when he decided to take his grandson along to share the hike, to share the stories.

He accepted an invitation to come to my lecture that night on the influence of African-American voices on Twain's work, and the role of Mary Ann Cord in particular, which was being given at Quarry Farm. It was a special pleasure to introduce him as an Elmiran with a family connection to a storyteller who had helped change the shape of American literature. After the talk, before he had to return to town, we repaired to Mary Ann Cord's kitchen, where we continued our conversation.

*   *   *

September 14, 1995. I sat in the rocker on Mark Twain's porch one last time sipping my morning coffee. As I looked out at the rolling lawn, the valley, and the hills, I reflected that there were other aspects of Quarry Farm that may have helped spark recollections of Mark Twain's childhood.

The group of children who played on the farm after the birth of Susy Clemens in 1872 and Clara Clemens in 1874 was a multiracial one. Louise Condol was a few years older than Susy. She spent a good part of her childhood summers visiting her grandmother when Mark Twain and his family lived at Quarry Farm, and remembered playing with the Clemens children and listening to Mark Twain read. Another black youngster whose name was also Susy—the child of tenant farmer John T. Lewis— was often at Quarry Farm as well. As a boy Twain himself had been part of a multiracial peer group in Hannibal, as is apparent not only from his fictionalized versions of it—such as the "pump" scene in *Tom Sawyer*— but also in his schoolmate John Ayres' recollections. And, of course, during summers on the Quarles' farm in Florida, Missouri, Clemens' playmates included a number of slave children as well as his cousins and siblings. The relaxed atmosphere Susan Crane created at Quarry Farm— she sometimes called the place "Go-as-You-Please Hall"—allowed the children to indulge in high-spirited, unfettered play that may well have reminded Mark Twain of scenes from his past.

The physical resemblance of Elmira to Hannibal—from a child's-eye view, at least—may also have made it easy for Twain's imagination to move easily from one to the other. Donald Blandford recalled that when he first read *Tom Sawyer*, he thought the book was about *him*. He and his pals in Elmira had woods, hills, caves, a river—just like Tom and Huck. They found uses for each not unlike those the boys found in Twain's novel. (Sometimes the hills and woods he explored as a child were actually at Quarry Farm, where Blandford's Boy Scout troop went camping.) James Lewis (interviewed in 1977 at the age of eighty-two), who used to play and hike around Quarry Farm, recalled "talking with Mark Twain on several occasions and telling him about a raft that he and some other boys made." Elmira, like Hannibal, was an outdoor paradise for a child. Watching from his porch or his study as black and white children darted in and out of the woods, ran up and down the hills, and played in the river, how could Twain *not* find the Matter of Hannibal close at hand? How could he *not* be drawn to recreate the exhilarating freedom of childhood so vividly embodied in the scenes around him? (Was the

9. *Mark Twain's study at Quarry Farm in the wooded hills overlooking the Chemung river (Photo courtesy of the Mark Twain House, Hartford, Connecticut)*

gloss of nostalgia that washed over his portrait of the violent and chaotic town of his youth in his first novel set in that town the product, in part, of his viewing the past through the lens of the comparatively peaceful and serene Elmira present?)

* * *

In June 1876 Mark Twain published *Tom Sawyer*, the book that helped earn "an almost mythical status for the Matter of Hannibal," as Henry Nash Smith put it, and "made Mark Twain's boyhood an international possession." While some critics argue that Tom's mild rebellions against social institutions like church and school prefigure themes of *Huckleberry Finn*, most continue to find *Tom Sawyer* a nostalgic paean to boyhood innocence and the genial virtues of small-town life in America, a story more evocative of "Old Times on the Mississippi" (published a year earlier) than indicative of things to come. In the 1930s *Tom Sawyer* was adapted to film twice, testimony to Hollywood's attraction, at the height of the Great Depression, to the book's "pre-industrial pieties." Interestingly, Gary Scharnhorst suggests that David O. Selznick's 1938 film *The Adventures of Tom Sawyer* was a "visual rehearsal" for a movie released by Selznick's studio the following year, which, like *Tom Sawyer*, was filmed in Technicolor, valorized the antebellum South, and was cast only after a nationwide talent search. That movie was *Gone With the Wind*.

In both *Tom Sawyer* and *Gone With the Wind* slavery is the backdrop rather than the focus of the action, the setting rather than the center of interest. None of the main characters in either enterprise is black, the black characters who *are* present play largely comic or stereotypical roles, and there is no one even remotely resembling a Mary Ann Cord. The racial hierarchies of the community—like the racist attitudes of its inhabitants—are simply assumed. The kind of moral indignation Twain had been cultivating in the 1870s is nowhere to be seen. Absent in *Tom Sawyer* is the ethical outrage that fueled "Only a Nigger."

Mark Twain's " 'hymn' to boyhood" (as *Tom Sawyer* has come to be known) would seem to have little in common with "Facts Concerning the Recent Carnival of Crime in Connecticut"—his dark, surreal tale of a man who kills his conscience—which appeared in June 1876, the same month he published the novel. On some level, however, Twain may have suspected that to recreate the boyhood pastoral of *Tom Sawyer* effectively, he had had to suppress that troublesome thing called a "conscience" that had begun to make him ask some difficult questions—such as whether that boyhood world was not so "innocent" after all.

In a letter he wrote at the end of August 1876, Mark Twain tore into his old friend Will Bowen for not relegating "the dreaminess, the melancholy, the romance, the heroics" of adolescence to the rubbish heap. "All this is," Twain ranted, "is simply mental & moral masturbation" belonging "eminently to the period usually devoted to PHYSICAL masturbation, & should be left there & outgrown." An odd comment, perhaps, from the author of *Tom Sawyer*, in which the dreaminess, melancholy, romance, and heroics of youth played a central if not defining role. At the time of his letter to Bowen, Twain had embarked on a new project and was striving for a tone as different from that of *Tom Sawyer* as "Disgraceful Persecution of a Boy" was from "The Celebrated Jumping Frog of Calaveras County." He was still dredging up memories of his Hannibal childhood, but in the service of a very different end. In August 1876 his project had gotten off to a roaring start and then crashed to a sudden halt, perhaps inducing some of the frustration and bitterness that came through in the letter. "As to the past," Twain wrote Bowen, "there is but one good thing about it, & that is, that it IS the past—we don't have to see it again." The book that was giving him so much trouble was *Adventures of Huckleberry Finn*.

Twain had penned several hundred pages of a manuscript about a child who grows up in a world in which no one—including that child—questions the God-given legitimacy of the institution of slavery; in which everyone—including the well-meaning, church-going solid citizens—supports a system that is inhumane, illegitimate, and barbaric. By the end of August, Twain had the child narrator throw his lot in with a runaway slave and set off on a journey whose end, for the slave, was to be freedom. But the writing stalled, and Twain put the child and the slave ashore and shelved the manuscript. When Will Bowen's nostalgic and sentimental letter arrived, it was too handy a target. For in addition to being stumped about what to do with the characters he had brought into being, Twain probably didn't know what to do with his own memories either—particularly painful and uncomfortable images that this writing business must have brought to the surface. Given the changes in Twain's moral universe in recent years, his own failure, for the first thirty years of his life, to defy or even question the institution of slavery and the racism that supported it must have troubled him. Can you remake yourself? he must have wondered. Consider the following comments, which he wrote in a public letter to the editor of the *New York World* in 1877: "Where is the use in bothering what a man's character was ten years ago, anyway? . . . I do not value my character of ten years ago. I can go out any time and buy a better one for half it cost me. In truth, my character was simply in the course of construction then."

The Hannibal of Twain's youth, like the St. Petersburg of both *Tom Sawyer* and *Huckleberry Finn,* was a slaveholding society; but only in *Huckleberry Finn* would this fact struggle to the foreground. The world of childhood fantasy, play, and adventure had preoccupied him in *Tom Sawyer*; in *Huckleberry Finn* Twain became alternately fascinated, stymied, and inspired by a story in which an adult black male on a quest for freedom came to play a central role. Elmira once again may have provided some of the impetus for Twain's decision to make the theme of freedom and the barbarity of racism integral components of his fiction.

* * *

I followed directions to the public library, parked the car, and headed for the reference desk, where the librarian sent me to the microfilm copy of the local newspaper for the summer of 1880. I cranked quickly to July and then began to scan the pages more slowly. Yes.

Mark Twain spent July through October of 1880, a period in which he resumed work on *Huckleberry Finn*, after a hiatus of several years, in Elmira. During the last week in July, while his household must have been preoccupied with preparations for the imminent birth of Jean, Twain probably followed with interest the city's preparations for another imminent arrival—that of Frederick Douglass, characterized by the *Elmira Daily Advertiser* as "the most prominent and honored colored man in the world."

The black community in Elmira had long been vocal, active, and well organized, working with state and national groups to lobby politicians on a wide range of issues, and participating in various "colored conventions." Frederick Douglass had positive associations with the area: it had been the home of his friend Jervis Langdon and several fellow abolitionists; he had given a well-received speech there in 1848; and prominent local leaders of the black community like John Jones and John T. Lewis (figures who had become legendary among whites as well as blacks in the city) were delighted to welcome him. These factors may have shaped his choice of Elmira as the site where he would mark "the anniversary of British liberation of slaves and the Emancipation Proclamation of Abraham Lincoln, merged in one grand day of enjoyment and commemoration."

Even had Mark Twain not had an interest in Frederick Douglass, the speech Douglass gave on this occasion would have attracted his attention, with its references to the upcoming presidential election that fall, an election that interested him greatly. (During the final week of the campaign in October, Twain would give a lengthy pro-Garfield speech to an overflow crowd at a rally staged in the Hartford Opera House. He would give another political speech a few days later at a Republican victory celebration.) But Twain *did* have an interest in Douglass, who had become a personal friend, and in particular in Douglass as an orator. In 1869 he had written Olivia Langdon, then his fiancée, that "I would like to hear him make a speech."

Indeed, even with no interest in either the election or Douglass, it would have been impossible for Mark Twain to miss the noisy hoopla that engulfed Elmira on August 3, 1880. The event drew delegations from virtually every city and town within a hundred miles. Sixty-three guns were fired at 11 a.m. Well before the parade began, the "excitement reached the white folks, and the streets were thronged with expectant people." Music was provided by the La France Band, the Geddes Cornet Band, the Havana Cornet Band, and the Corning Band. Finally, "a drum corps of small boys brought up the rear." The procession's route was " 'fairly alive' with spectators, and the utmost enthusiasm and interest was manifested on all sides. There was music all along the route, in fact, with so many bands the air was 'full of it,' and no one could help being in sympathy with the occasion." The path of Twain's customary daily walk to town ran right into the parade route (which circled around the Langdon home at Church and Main streets); even if he had remained at Quarry Farm that morning, sounds of the brass bands and drums would have wafted up the hill, alerting him to the excitement sweeping the city.

Whether or not Twain attended the speech, which the *Advertiser* called "one of the strongest addresses ever heard in Elmira," the paper's detailed coverage of it would not have escaped him. In addition, Twain had daily conversations with John T. Lewis, one of the key players in all the Douglass-related festivities. Three years earlier Lewis had saved the lives of several members of the Langdon family by heroically stopping a runaway horse. Twain had great admiration for him and highly valued his friendship. As a member of the Committee on Arrangements and the Committee on Reception, Lewis played a leading role in planning and executing the day's events, the centerpiece of which was Douglass's speech, described glowingly in the *Advertiser*:

> As the venerated and noble colored man stood on the platform, with his head bared, his white and heavy locks, his massive frame and kindly eyes, gave him the appearance of a Moses of his race, and a man whom all would single out as a remarkable person, whose presence would be felt wherever he might be placed. . . . He was frequently interrupted with applause at the utterance of some sentiment, or laughter at some keen thrust at the enemy. We print his remarks on the eighth page, and recommend everyone to read them.

The *Advertiser*'s main interest in Douglass's oration was its political import. The paper was staunchly behind Garfield, as was Douglass; it viewed the speech as potentially useful in the coming election. "Although particularly intended for, and delivered to our colored citizens," Douglass's address, the paper declared, "should be carefully read and thoroughly perused by every voter, no matter what may be his nationality or the hue of his skin." To the *Advertiser*'s delight, Douglass's oration showed plainly "why Garfield and not Hancock should be elected." But he did much more than that.

Sandwiched between his political oratory and a discourse on the historical events the day commemorated were some hard-hitting attacks on what the nation had made of "freedom" as far as its black citizens were concerned:

> To-day, in all the Gulf States, the fourteenth and fifteenth amendments of the Constitution are practically of no force or effect. The sacred rights, which they solemnly guaranteed, are held in contempt and are literally stamped out in the face of the Government. By means of the shot gun and midnight raid, the old master class has triumphed over the newly enfranchised citizen and put the Constitution under their

feet. In South Carolina, Louisiana, and Mississippi, the colored people, who largely outnumber the whites, and who are Republican in politics, have been banished from the ballot box and robbed of representation in the councils of the nation, and according to the best information from that quarter the social conditions of the colored people in that section is but little above what it was in the time of slavery. In fact, the chain gang has re-appeared in those States, and persons of color, for the most petty offenses, are put in these gangs and made to work the farms of their former masters under the lash. There is but little trouble in convicting a negro in Southern courts, and in fact anywhere else.

The final portion of *Huckleberry Finn* is increasingly coming to be understood as a satirical indictment of the virtual re-enslavement of free blacks in the South during the 1880s. It is quite likely that Frederick Douglass's acidic characterization of this phenomenon influenced Twain. The trip south he would soon undertake to write *Life on the Mississippi* would further sharpen his understanding of these problems. He had penned satires about racism toward the Chinese, but with the exception of the antilynching editorial "Only a Nigger," and the brief comment on the incident involving the Buffalo police, he had not directly explored the contemporary problem of racism toward blacks. If Mary Ann Cord brought the 1840s more clearly into focus with her story of being pulled from her child on the auction block, Frederick Douglass forced Twain to confront the fact that the promise of freedom for black Americans in the South in 1880 was being flagrantly betrayed.

As Twain struggled to complete *Huckleberry Finn*, the issues Douglass raised would take on more and more importance. Elmira once again forced Twain to think about a troubling legacy of the Matter of Hannibal: the contemporary racism directed against blacks which, until *Huckleberry Finn*, Twain had still all but avoided addressing in his art. Elmira was far enough away psychologically and geographically from the terrain of his youth to give him the necessary distance, but it was close enough to embody vivid reminders of that world that would help him transmute it into fiction. Twain inscribed the set of his works that he presented to Charley and Ida Langdon in 1900 with the words "It is one's human environment that makes climate." Elmira proved to be a particularly fruitful climate for Twain to explore the Matter of Hannibal in his art.

*   *   *

Hartford, October 1, 1994. The burnished wood of the walnut banister reflected the light from the sconce on the wall in a lustrous glow. I had been in the Twain house many times before, but never after dark. It had been opened tonight for the participants in the Mark Twain House fall symposium and other special guests. Standing in the foyer of his home, I felt his presence strongly. The fixtures, designed to cast a light not unlike that which Twain and his family would have experienced after the sun had set, imbued the scene with a warm radiance conducive to flights of fancy. As I roamed from room to room, my mind wandered back and forth between meditations on Twain and recollections of the conversations I'd had earlier that day.

I walked through the brocade-upholstered drawing room and thought about the evening when Mark Twain appalled and astonished his uptight guests by giving a little demonstration of the nineteenth-century precursor to break dancing. It was a harbinger of things to come, I thought, as I recalled the confusion he wrought by making great literature out of voices belonging to slaves and servants and whites like Huck at the bottom of the heap. It was here, I thought, in this house, that Twain was living when the story Mary Ann Cord told him appeared in *The Atlantic Monthly*, when the *New York Times* published the piece about the black child named Jimmy who entranced him during his Midwestern tour, and who would become a key model for the voice he would later give Huck Finn. It was here where he made the decision to underwrite the apprenticeship of Charles Ethan Porter, a black painter who wanted to study in Paris. And where he signed the check that helped put A.W. Jones, a black theology student, through Lincoln University, and where he said yes to invitations to give readings in local black churches to help them raise funds.

This was the house that George ran—George Griffin, known sometimes as Twain's butler and sometimes as his "liar" (he was charged with quickly dispatching unwanted guests), but who might best be described as his friend. The smart, savvy black man, a former slave, had shown up one day to wash windows, and ended up moving in. He remained for years and stayed in touch with Twain and his family after they left Hartford. In the plush dining room with the fireplace with the artfully divided flue, when Twain held forth with jokes, George would often smell the punch line a mile away and burst into peals of laughter long before Twain got to it. Stories about George Griffin had long been a part of the tour guides' commentaries as they took groups through the house, but the

10. *The Mark Twain House at 351 Farmington Avenue in Hartford, Connecticut,*
where Twain lived from 1874-1891 (Photo courtesy of the Mark Twain House,
Hartford, Connecticut)

room George had lived in (down the hall from Twain's billiard room)
had not been a part of the tour. I was pleased to learn that funds were
being raised to restore and refurbish it. Work was scheduled to begin
later in the year. As I walked back through the foyer, I tried to imagine
George gregariously pumping Twain's political pals for news as he
greeted them and took their hats (armed with the inside dope, George
would take election bets in the black community and clean up).

The image of George Griffin collecting visitors' hats reminded me of
stories I'd read about a young black man named Warner McGuinn who
had worked his way through Lincoln University by spending summers
manning the hat check counter at the Newport Casino in Rhode Island.
He was renowned for being able to take the hats of four hundred men,
note their faces and the shape of their heads, and return the right hat to
each man without ever getting it wrong. He had been a major talent as a
hat check clerk, no doubt about it. But he was destined for better things.

I wandered upstairs to the billiard room, where Twain did most of his writing. Was it here, on Christmas eve in 1885, that he wrote the letter to the dean of the Yale Law School that changed Warner McGuinn's life and my own?

*   *   *

New Haven, February 4, 1985. My husband roused me from the deep sleep known to mothers of active two-year-olds. "Wake up," he said. "You've got to hear this!" I looked out the window. It was dark. I glanced at the clock. It was a little after eleven.

I followed him into the living room. On ABC's *Nightline* Dr. John Wallace, a Chicago-based educator, was shouting that *Huckleberry Finn* was "the most grotesque example of racist trash ever written." His immediate objective was to close down the stage production of it at Chicago's Goodman Theater before any more damage was done. A black Chicago alderman sitting next to him seconded the motion. On the opposing side were civil libertarian Nat Hentoff and Meshach Taylor, the actor playing Jim at the Goodman. Ted Koppel asked Meshach Taylor whether he had had to do some "soul-searching" before agreeing to play Jim, given all the controversy surrounding the book. "Initially I did," Taylor said. When he considered appearing in a dramatic adaptation of the book for the first time ten years earlier, he had never read the book, but he had seen "old movies" of it. The director insisted that he read Twain's novel, which he did.

> [I] felt it was one of the best indictments against racism in the United States that I ever read. And the character of Jim appealed to me because I felt that an individual who speaks with a dialect is not necessarily an ignorant individual. It's an individual who may be uneducated but not stupid.

At this point Koppel interjected, "And you do not feel that it enhances bigotry when words like 'nigger' are used on the stage, that this somehow gives the bigots out there perhaps a feeling that it's all right?" Taylor responded,

> No. First of all, let me say that I feel that the word "nigger" is an offensive word. I do feel that slavery was offensive as well. I think that if we're going to be true to the time, however, that we must speak the way the people spoke during that time. And I think it's important for people to understand exactly what the history of racism in this country is.

John Wallace was unmoved. "I think the book is certainly the most

racist book, among many, that is printed in the United States of America."
Neither Hentoff nor Taylor is ever short on opinions or shy about expressing himself, but pretty soon they both seemed to be fighting a losing battle. No sooner would one of them open his mouth than Dr. Wallace would cut him off and launch into a variation on his favorite theme: Twain was a racist author, *Huckleberry Finn* was a racist book, and neither had a place in the classroom.

I was up all night, pouring my anger into my word processor. The result was a column that the *New York Times* published on its opinion page on February 18, 1985, the centennial of the U.S. publication of *Huckleberry Finn*. I argued that it was doubly ironic that Mark Twain, who had turned to irony in the first place to avoid being censored in his attacks on racism, should now be attacked as a racist himself by people who missed the point of his irony. The day the essay ran in the paper, I was awakened at 6:30 a.m. by a phone call from an antiques dealer in Hamden, Connecticut. "You don't know me," Nancy Stiner said, "but I just read your piece in the *New York Times* and I've got to see you right away. I have a letter Mark Twain wrote that nobody knows about yet, and after reading your column I know you'll know what to do with it. Here's what it says." She read me the letter over the phone.

December 24, 1885

Dear Sir,
Do you know him? And is he worthy? I do not believe I would very cheerfully help a white student who would ask a benevolence of a stranger, but I do not feel so about the other color. We have ground the manhood out of them, & the shame is ours, not theirs, & we should pay for it.

If this young man lives as economically as it is & should be the pride of one to do who is straitened, I would like to know what the cost is, so that I may send 6, 12, or 24 months' board, as the size of the bill may determine.

You see he refers to you, or I would not venture to intrude.
Truly yours,
S. L. Clemens.

A chill went through me as I realized the significance of what I had just heard. The letter contained the most direct, non-ironic condemnation of racism that we had from Twain himself during this period—and it was written in Hartford on Christmas Eve 1885, the same year *Huckleberry Finn* was published.

"Where did you find it?" I asked.

"In an old desk."

"I'll be right there."

She lived fifteen minutes away. I dressed and drove out to her house. She had had to leave for work, but her husband was waiting for me with the letter. The handwriting was unmistakably familiar, but how could I be sure?

Nancy and Richard Stiner let me borrow the letter to compare it with other Twain letters in Yale's Beinecke Library, many of which were written from Hartford around the same time. I made a photocopy and resolved to determine the circumstances under which the letter was written—to whom and about whom—and what happened after it was sent.

I called the Stiners, told them I was fairly certain it was the real thing, and arranged to go out to their home to return the original. A torrential rain pounded the window of my New Haven apartment as I looked at the glass-framed letter one last time before wrapping it in double layers of cloth and plastic and placing it in my briefcase as gently and carefully as I would have placed a bomb; indeed, in some ways that's exactly what it was. I smiled when I thought of how this letter would take the wind out of the sails of those who wanted to kick Twain out of school as a racist—*if* it was authentic.

That was the biggest question on my mind. What if this were a forgery? If it were, wouldn't Nancy Stiner have to be in on it? Of course, I didn't know her very well, but she struck me as too sensible to do anything like that. In fact, I was sure of it. Still, there was always that little shred of doubt.

The rain pelted the windshield relentlessly as I drove up her driveway. I wrapped more plastic around my briefcase, opened my umbrella, and made a dash for the back door of the house, where the tall, silver-haired antiques dealer was waiting for me. We chatted for a few minutes, and then I returned the letter, thanking her for having let me borrow it.

"Oh, by the way," she said as I was leaving, "I don't think I told you that my father was Mark Twain."

My heart sank. If someone was capable of cooking up the fantasy that she was Mark Twain's daughter, was there any limit to what else she might do? Nancy Stiner gave me a puzzled look, wondering why I looked so unhappy all of a sudden. Crazy people are like that, I thought. They never realize the effect they have.

"Your father was Mark Twain," I repeated slowly, conscious of the feeling of sawdust in my throat.

"Yes," Nancy Stiner said brightly. "Every year for years. In the Memorial Day Parade in Hartford. He was quite wonderful at it."

* * *

The reply to Twain's letter—which was in his papers at Berkeley—and an article by literary scholar Philip Butcher made it clear that the addressee was Francis Wayland, dean of the Yale Law School (the desk in which the letter was found had belonged to one of Dean Wayland's descendants), and that "young man" to whom Twain referred was Warner T. McGuinn, one of the law school's first black students. But McGuinn himself was still largely a mystery.

Most of the law school's nineteenth-century records had been put out on the curb years before. Sitting in the Yale Archives room in Sterling Memorial Library, I devoured whatever I could find. A handful of registers and rosters. Some committee-meeting minutes. A record book of awards and prizes. I heard lights flipping off around me. Then I saw his name under the listings for 1887, the year he graduated: Warner McGuinn, Townsend Prize. "I'm sorry," said the librarian. "We're closing." He turned off my light. Damn. I made note of the page number, returned the book to the reserve cart, and walked out through the thick double Gothic doors to a now dark Wall Street, planning my next step. The university registers. Try to find out where he lived. I did a U-turn at my front door. I had forgotten to pick up the milk we needed for breakfast. As the cold winter air cut through my thin sweater, I realized that I had left my coat on the library coat rack. I took extra care crossing the street: I knew that the present had pretty tough competition for my attention. I was deliciously, dangerously possessed by this fragment of the past.

I returned to the library the next day and every day after that for weeks, following McGuinn's elusive trail: a scrapbook he kept that someone had sent to the library after his death but no one had ever looked at before; a microfilm copy of the *American Citizen* from Kansas City, Kansas, the newspaper he edited after graduation; his 1937 obituary in the *Baltimore Afro-American*, which revealed the contours of a career as distinguished as it was obscure. Born near Richmond, Virginia, in 1862, McGuinn attended Lincoln University in Pennsylvania and then read law in Washington, D.C., with Richard Greener, the first black graduate of Harvard. In 1885 McGuinn was admitted to study law at Yale. He supported himself as a waiter, accountant, and bill collector during his first year in New Haven. The one extracurricular activity for which he managed to find time was the Kent Club, a student organization that debated political and social questions of the day. Two warring factions vied for the presidency of the club in the fall of 1885, and McGuinn was finally selected as a compromise. One of his duties was to meet the club's guest speakers at the railroad station and introduce them before their lectures. The first speaker that year was Mark Twain.

McGuinn met Twain at the station, escorted him to the dean's house, and later introduced him at a public meeting. Twain must have been impressed by the young law student and by the introduction he gave. Back in Hartford, Twain wrote Dean Wayland on Christmas Eve, inquiring about McGuinn's circumstances and asking whether it was a good idea for him to offer to pay his board. Wayland responded on Christmas Day: "I think the colored youth is a promising case & deserving of help from someone." Five days later the dean wrote Twain again, adding that McGuinn was "very studious, & well behaved, above the middle of a very good class & very anxious to succeed." Twain paid his board for the next year and a half (the remainder of McGuinn's time at Yale), receiving interim reports from Wayland, like the one dated October 1886 in which he called McGuinn "the most promising colored youth we have ever had."

The final assignment required for graduation was to write either a thesis or a Townsend Oration. Law school regulations specified that Townsend Orations were to be twelve to eighteen pages long, on one of four possible topics. McGuinn chose the Townsend Oration option and the topic "The Constitutional Limitations to Land Taxes." In this blind competition, in which entries submitted under fictitious names were evaluated by a committee of distinguished judges, McGuinn came out on top, winning a prize of one hundred dollars. Dean Wayland wrote Twain in the fall of 1887:

> Your beneficiary of last year & year before—a graduate of Lincoln University—made a very creditable record for himself in this school & took the prize for the graduating oration by award of the very competent committee, consisting of an ex-Minister to England, a Judge of the Supreme Court of the U.S. & one of the leading members of the Conn bar.

After serving as editor of *The American Citizen* in Kansas City, McGuinn set up a law practice in Baltimore, where he won a major civil rights victory in federal court, helped found the local branch of the NAACP, served as counsel to the *Baltimore Afro-American*, was elected to the city council, and became a mentor to a young attorney in the office next door named Thurgood Marshall.

Reading Warner McGuinn's scrapbook was a disorienting experience. Like a curiously postmodern palimpsest, it was full of pages torn from a volume entitled *Vital Statistics*, on top of which were pasted newspaper clippings from both mainstream and African-American newspapers. Mortality and disease statistics in specific wards and districts of Baltimore and D.C., broken down into"white" and "colored" columns, framed stories about disenfranchisement, discrimination, segregation, and racially motivated violence. McGuinn had run precisely the same kinds of stories when he edited *The American Citizen* some thirty years earlier. All this material documented in starkly vivid terms chapters of American history only vaguely familiar to me from standard textbooks. Under McGuinn's tutelage, I found new dimensions of the past coming alive. The articles limned the period known as the "Nadir," the all-time low point in American race relations (1877 through about 1915) and the years that followed. One article described a near-riot in New Orleans provoked by black actor Ira Aldridge's masterful stage portrayal of Othello. A letter to the editor proposed that the "constantly increasing" class of "negroes, especially those in the cities . . . men who idle away their lives," be conscripted to "make our proposed army" and thereby "pretty nearly settle the so-called negro question." Another article quoted a local politician's warning: "Once the negroes began to join, it was declared, the Southern white boys would stay at home." Yet another clipping reported the cheers that greeted the introduction of two "Jim Crow car" bills in the Maryland legislature.

Above two dense columns in the scrapbook headed "city ordinances," McGuinn had scrawled "Segregation Law" in thick black letters. A few pages later I saw why those columns had been so carefully clipped and preserved. In federal court in the autumn of 1917 McGuinn challenged that law. Maryland's white establishment lined up against him, confident of their ability to maintain the status quo (the opposing team included the attorney general, the deputy states attorney, and the city solicitor of Baltimore). According to newspaper reports, McGuinn thoroughly bested his opponents. As his obituary in the *Baltimore Afro-American* described it:

> Arguing the celebrated Baltimore segregation case in 1917, before Judge John C. Rose in the Federal Court, Mr. McGuinn objected frequently to the points of law made by opposing counsel and finally asked the court to have him argue the case at issue.

Judge Rose in a squeaky voice replied:

"Leave him alone, Mr. McGuinn, he's doing the best he can."

When the case was decided, the *Philadelphia Tribune* ran McGuinn's picture under the headline THE ATTORNEY WHO KNOCKED OUT THE BALTIMORE SEGREGATION ORDINANCE. His victory discouraged Baltimore and other Southern cities from passing further segregation ordinances and played a major role in the eventual legal (if not de facto) desegregation of American cities.

During the weeks after the Twain letter came into my life, Warner McGuinn—and the history he preserved and made—had captured my attention more than any research I had pursued in a long time. In 1985 there was no recognition of his achievements or even official awareness of his existence at Yale, the university that had trained both of us in our chosen professions. I hoped my research would help change that, and I'm gratified that to some extent it did (his picture now hangs in the law school). I also hoped that the recovery of the Twain letter would clarify for the public Twain's views on racism and its legacies at the time of *Huckleberry Finn*, and thus act as a counterweight to the refusal in some quarters to acknowledge the irony that characterized his published writing on race from that period.

Contacted by Yale's news bureau, the *New York Times* expressed interest in running the story, but their reporter, Edwin McDowell, wanted to reach U.S. Supreme Court Justice Thurgood Marshall first. Justice Marshall was no fan of Mark Twain's and was somewhat unclear about why the *Times* was eager to talk with him, but when he finally understood that the subject was Warner McGuinn, McDowell got his quote.

> "He was one of the greatest lawyers who ever lived," said Thurgood Marshall, Associate Justice of the Supreme Court of the United States. Justice Marshall, who as a young lawyer in Baltimore shared adjoining offices with McGuinn, said yesterday that McGuinn enjoyed the respect of the entire bar and judiciary. "If he had been white, he'd have been a judge," Justice Marshall said.

I was stunned, on the morning of March 14, 1985, to see McDowell's story occupying nearly the entire lower-left quadrant of page 1.

Twain's condemnation of racism in the 1880s was not news to scholars, who had long been familiar with his private, unpublished jibes at it from the 1870s on. Nor was it news to careful readers of *Huckleberry Finn*, who grasped the antiracist thrust of the book's satire. But Twain's irony was hard for some to recognize, and the book was filled with a racial epithet that retained its power to inflict pain. Many ordinary readers simply believed John Wallace when he charged that Mark Twain and his book were racist, pure and simple. The letter provided Twain's defenders with welcome ammunition.

"It's the smoking gun. The man was not a racist," said the renowned black scholar Henry Louis Gates Jr. on national television that evening. I gave interviews to the *Today Show,* the *ABC Evening News,* a dozen radio stations across the country, a newspaper in Australia (which ran the headline TWAIN'S BLACK MARK REMOVED), a radio station in Canada, and a literary publication in Moscow. I also debated John Wallace on CNN that night and the next morning on the *CBS Morning News.*

The letter clearly made Dr. Wallace unhappy. He was at a bit of a loss about what to say. The best he could muster was that Twain should have paid for a whole group of black students at Yale, not just one. He repeated the charges he had made on *Nightline*—that the book was racist, its author was racist, and it had no place in America's classrooms. I didn't doubt that Twain's novel had been poorly taught in the school Dr. Wallace attended, and that it had caused him pain and discomfort as a result. But it seemed to me that the solution was to show teachers who wanted to teach the book how to teach it more successfully—how to frame the discussion in the context of American race relations and help students recognize irony—rather than force them to remove it from their reading lists. During the call-in portion of our hourlong debate on CNN, a black high school student accused Dr. Wallace of insulting her and all black high school students by suggesting they weren't smart enough to understand Mark Twain's irony. I felt like giving her a hug.

I called my father after the show. "How was I?" I asked. He paused and then said, "Too ladylike." He was right. Dr. Wallace had interrupted and harangued me at every turn. I had been too ladylike to do anything about it.

I knew I'd have a rematch, albeit a brief one, on tomorrow's *CBS Morning News.* I woke up two hours earlier than I had to just to get psyched. Before the broadcast Dr. Wallace and I chatted casually over coffee. We got along fine when we weren't talking about Mark Twain. He seemed relaxed when we entered the set: he had me pegged, having debated me the previous evening. Only the "me" who turned up that morning wasn't quite the same "me" of the night before. His "ladylike" opponent was gone, and in her place was someone who held her ground, ignored his interruptions, and finished her sentences against all odds.

When the show was over, a black cameraman who had been shooting us came up to me. He said he had never read the book by Mark Twain that we had been arguing about but now he really wanted to. One thing that puzzled him, though, was why a white woman was defending it and a black man was attacking it, because as far as he could see, from what we'd been saying, the book made whites look pretty bad.

Back home letters and phone calls began pouring in, not all of them from fans. There was the postcard from the woman who was irate about my assertion that Twain was not a racist: of course he was a racist, she protested, and that was one of the things she liked about him. And there was the time my husband answered the phone and a voice demanded, "Are you the feller married to that nigger-lover?"

*   *   *

Hartford, October 1, 1994. The Mark Twain House symposium had brought the pioneering comic Dick Gregory, cultural critic Michael Eric Dyson, columnist Clarence Page, novelist Gloria Naylor, journalist Andrea Ford, folklorist Roger Abrahams, myself, and others to town for a day of panel discussions, followed by dinner and a tour of Twain's house. The symposium had originally been called " 'Nigger' and the Power of Language," a title that quickly proved too combustible; the initial epithet was dropped from all but the most preliminary advance notices about the event. But the word had hovered behind the day's discussions, whether I was quoting Twain's 1869 antilynching editorial "Only a Nigger" or whether Dick Gregory, who had titled his autobiography *Nigger*, was holding forth on Twain's genius as a satirist.

I had reread Gregory's book on the plane to Hartford and was struck anew by its brashness and bite, starting with the dedication on the book's first page:

Dear Momma—Wherever you are, if you ever hear the word "nigger" again, remember they are advertising my book.

When I read Gregory's masterful jabs at Southern good ol' boys, I recalled Twain's comment about no tyranny being so strong that it can withstand the weapon of ridicule:

"Last time I was down South I walked into this restaurant, and this white waitress came up to me and said: 'We don't serve colored people here.' "

"I said: 'That's all right, I don't eat colored people. Bring me a whole fried chicken.' "

"About that time these three cousins come in, you know the ones I mean, Klu, Kluck, and Klan, and they say: 'Boy, we're givin you fair warnin'. Anything you do to that chicken, we're gonna do to you.' About then the waitress brought me my chicken. 'Remember, boy, anything you do to that chicken, we're gonna do to you.' So I put down my knife and fork, and picked up that chicken, and kissed it."

And I was moved by the way Gregory combined bold social critique and a lyrical paean to

all those Negro mothers who gave their kids the strength to go on, to take that thimble to the well while the whites were taking buckets. Those of us who weren't destroyed got stronger, got callusses on our souls. And now we're ready to change a system, a system where a white man can destroy a black man with a single word. Nigger.

The book was first published in 1964—soon after the bombing of the Birmingham church that killed four little girls and the assassination of Medgar Evers, and shortly before Watts, the murders of the three civil rights workers in Mississippi, and my first encounter with Pap Finn in high school. Gregory's narrative of his experiences brought back to me the violent lunacies of that era more vividly than anything I had read since. As I reread the book, something else caught my eye that had somehow slipped my attention the first time around: Gregory hailed from the same home state as Mark Twain.

Both of us were scheduled to appear on the morning panel. The van from the hotel dropped all of us off at the Aetna Center. We were instructed to make our way to the Green Room until it was time to go on stage. Dick Gregory had been a hero of mine for so long that it was with a rather starstruck awe that I found myself at his side as we headed down the long corridor.

I nearly tripped when he told me he was aware of my work. (My book *Was Huck Black? Mark Twain and African-American Voices* had come out the year before).

"Hm-hm," he said. "We always said Mark Twain got his stuff from things black folks told him. My *grandmother* said that. Now white folks with Ph.D.s get government grants to discover that stuff. . . ." He shook his head.

I started to explain that I hadn't ever gotten any government grants to discover anything, when I stopped myself. That wasn't really the point. Gregory was right. Blacks had known it all along. True, they hadn't "proved" it. But they had known it and said it. Only nobody out there listened. Well, almost nobody.

Nine years earlier, in the spring of 1985, shortly after my adventures with the McGuinn letter, the Mark Twain House (then called Mark Twain Memorial) invited me to be a guest at their annual meeting, an event held jointly with the New England American Studies Association convention. The award-winning novelist David Bradley, author of *The Chaneysville Incident* (my candidate for the great American novel of our time) was scheduled to speak. He titled his talk provocatively, "The First 'Nigger' Novel."

Bradley stared out into the crowd of mainly white American Studies scholars and Hartford patrons of the arts and said, "You folks know a lot about Sam Clemens. Sam Clemens was white. But who here among you has ever seen Mark Twain? Mark Twain was black." He then proceeded to make a case for *Huckleberry Finn* as a work which prefigured the fiction of African-American writers in the twentieth century—including his own. The audience, to put it mildly, was in shock. Some were outraged. Some felt threatened. Others were simply confused.

I remember clearly my own response: he was right. His speech had been a virtuoso performance rather than a scholarly disquisition. But his insights were right on target. It made perfect sense to view *Huckleberry Finn* as a key precursor to a great deal of fiction by black writers that came after it, just as he said. I was certain of it. I had no more proof at that point than he did, which, save for when he was speaking as a writer himself, was none at all. Could I prove it? I wondered. Slave narratives had rarely employed dialect, instead seeking to demonstrate through well-crafted, Standard English prose the ex-slave's claim to a place at the table of humanity. And most of Twain's black contemporaries (with one or two exceptions) had steered clear of using the vernacular in their work as well, preferring the measured tones of the educated middle class instead. It was certainly plausible that Twain had been an important influence on writers like Langston Hughes and Ralph Ellison. And hadn't Bradley acknowledged his own debt to Twain?

The seeds that were planted that night took six or seven years to germinate. During that time, as my antennae picked up everything that had bearing on the subject, I found a paper trail that supported Bradley's argument. Black writers who admired Twain included Charles Chesnutt, who kept a bust of Twain in his library, Ralph Ellison, who kept a photo of him over his desk, and Langston Hughes and Richard Wright, both of whom had paid eloquent homage to Twain in print. Through conversations and correspondence over the next few years, I found that Twain had been important to other contemporary black writers besides Bradley, including Toni Morrison, who returned to Twain when she was honing her craft as a writer. It was during an interview with Ralph Ellison in 1991 that my own variation on Bradley's theory began to take shape. Ellison spoke of Twain's special appreciation of the vernacular and of the irony at the core of a nation founded on ideals of freedom that tolerated slavery and racism in its midst. Mark Twain, Ellison said, "made it possible for many of us to find our own voices." *Why* had Twain played this empowering role for black writers? I wondered aloud. Could some of the things Ellison learned from Twain be things Twain himself had learned from the rhetorical performances of African-Americans? Yes, Ellison responded, "I think it comes full circle."

From that moment on, I began to systematically track all black speakers in Mark Twain's work. I reread a posthumously published essay by Twain in which he referred to an "impudent and delightful and satirical young black man, a slave" named Jerry (whom he recalled from his Missouri childhood) as "the greatest orator in the United States." I also reread an obscure article Twain had written in the *New York Times* in 1874 about a ten-year-old black child named Jimmy who had impressed him as "the most artless, sociable, exhaustless talker" he had ever come across, someone to whom he had listened "as one who receives a revelation." I found compelling evidence that black speakers had played a central role in the genesis not only of Twain's black characters but of his most famous white one: Huckleberry Finn.

If black oral traditions and vernacular speech had played such an important role in shaping Twain's art, why hadn't anyone noticed it before, given the thousands of books and articles on Twain that had appeared? Literary scholars had denied any African-American influence on mainstream American texts, much as linguists had denied any African-American influence on Southern speech and American speech in general. All of them, I became increasingly convinced, were wrong.

At the Mark Twain Papers at Berkeley, I examined, among other things, the manuscript of that essay about the "satirical young black man" named Jerry, and found that Twain had first called him "the greatest *man* in the United States." Back in Austin, I mined published and unpublished fiction and nonfiction by Twain, folklore and linguistic studies, history, newspapers, letters, manuscripts, and journals. I didn't come up for air all spring.

As I knew from my first encounter with the book in high school, critics had long viewed *Huckleberry Finn* as a declaration of independence from the genteel English novel tradition. Something new happened here that had never happened in American literature before. *Huckleberry Finn* allowed a different kind of writing to happen: a clean, crisp, no-nonsense, earthy vernacular kind of writing that jumped off the printed page with unprecedented immediacy and energy; it was a book that talked. I now realized that despite the fact that they had been largely ignored by white critics for the last hundred years, African-American speakers, language, and rhetorical traditions had played a crucial role in making that novel what it was.

Ralph Ellison had Mark Twain's number. He had written that "the spoken idiom of American negroes [was] absorbed by the creators of our great nineteenth-century literature even when the majority of blacks were still enslaved. Mark Twain celebrated it in the prose of *Huckleberry Finn*." But his comment sank like a stone. "The black man," Ellison had said, was the "co-creator of the language Mark Twain raised to the level of literary eloquence." But literary historians ignored him and continued to tell us that white writers came from white literary ancestors and black writers from black ones. I knew that story had to change if we wanted to do justice to the richness of our culture.

I hadn't dialed his number in years, but I knew I had to call Bradley and tell him. After all, wasn't it really his idea? I called him in February 1992. "This may sound crazy," I remember saying, "but I think I've figured out—and can prove—that black speakers and oral traditions played an absolutely central role in the genesis of *Huckleberry Finn*. Twain couldn't have *written* the book without them. And hey, if Hemingway's right about all modern American literature coming from *Huck Finn*, then all modern American literature comes from those black voices as well. And as Ralph Ellison said when I interviewed him last summer, it all comes full circle because *Huck Finn* helps spark so much work by black writers in the twentieth century."

I stopped to catch my breath. There was a pause on the other end of the line. Then a question: "Shelley, tell me one thing. Do you have tenure?"

Ralph Ellison had been kind enough to read my manuscript through before I went public with my findings. When we spoke on the phone after he read it, he couldn't help chuckling with pleasure, delighted to have his intuitions validated after all these years. Hal Holbrook, who has embodied Mark Twain's voice for four decades, also read the manuscript. He told the *San Francisco Chronicle* that he had "sensed a black strain in Huck's voice but never knew for sure," adding, "It's almost like the truth about something is so clear that you can look right through it."

Senior scholars who had devoted their lives to Twain's work were, for the most part, wonderfully open to my findings and remarkably supportive. The occasional curmudgeon, of course, willfully ignored the fact that I was talking about the genesis of Huck's *voice* and not his skin color. And a scholar who should have known better argued something to this effect: How can Huck's voice be black if a sizable portion of it comes from white humorists? I had assumed that all of my readers would be able to recognize that my book's title was "signifying" on the famous "one-drop rule" that legally defined a person as "black" if he or she had only the most minute amount of "black blood." In retrospect I realize that was a mistake. The concept—crucial as it is to understanding the history of American race relations—had not made it into either E. D. Hirsh Jr.'s list of "What Every American Needs to Know" or the general "cultural literacy" of the country. I should have explained it: Huck's voice didn't have to be "*all*-black" in order to be considered "black," according to the traditional law of the land, it only had to be "part-black." And the evidence that his voice had been shaped at least in part by African-American voices was strong, as even the most skeptical critic had to concede.

*   *   *

Standing in the foyer of Mark Twain's house, waiting for the rest of the group to join me to take the bus back to our hotel, I wondered how many of us had been drawn to visit this lavish, sophisticated mansion because the man who had lived there had painted a simple dawn on a silent river that was more real and memorable than any dawn we had ever witnessed? I thought about the demands Twain put on us, the contradictions he required us to acknowledge and address. The paterfamilias hosting elegant dinners in this house in Hartford also contained within himself the unruly child who hated to put on shoes. The man who felt such a deep sense of shame about the role white people played in oppressing blacks in America that he made that oppression central to his greatest works of fiction, explored the subject so artfully that he would be constantly misunderstood. Why was he so cagey? Why so reticent to stake out these positions unequivocally? I thought of the numerous fragments Twain wrote but chose not to publish. So many—like "The United State of Lyncherdom"—dealt with issues of race. Twain wrote his publisher that he would not have a friend left in the South if he went through with that book. (He seems to have been unaware of the fact that black writer Ida B. Wells had written a very similar book several years before Twain conceived his). Was Twain guilty of trying to "have it all," to be true to his principles yet retain that "option of deniability" that could banish controversy from his doorstep when he chose to do so? Did the local-boy-made-good who relished the chance to "go home again" to a hero's welcome somehow make it too easy for his countrymen to avoid confronting the dark currents under the raft?

As I stepped out into the crisp night air of a New England autumn, I thought about how many Twains there seemed to be—and how the ones we choose to make our own reveal us to ourselves in fresh and surprising ways.

*   *   *

Dallas, October 10, 1995. Texas state senator Royce West ushered me into his bright, well-appointed law office. I had come to town the night before to lecture under the auspices of the African-American Museum, an impressive institution that had been the beneficiary of Senator West's enthusiastic support. I had admired the senator's past initiatives in such areas as juvenile-justice reform and was pleased when he agreed to talk with me about a puzzling piece of legislation he had introduced the previous spring.

Senator West's amendment to an amendment to the education bill had stipulated that "each approved textbook must be free of language that refers to one or more specific racial groups and that, in the opinion of the [school] board, considering the context in which the language occurs, is derogatory or malicious and has no education purpose [sic] appropriate to the age of the student." No state funds could be spent to purchase any book that failed to meet these criteria. When the press referred to the amendment as the "*Huck Finn* bill," I decided there had been some mistake; the loose language of the bill must have accidentally caught Twain in its net, I thought. I was wrong. It was called the "*Huck Finn* bill" for a good reason. It targeted *Huckleberry Finn*. Senator West wanted the novel out of Texas classrooms.

"Could you tell me a little bit about how the amendment to the Nelson amendment came to be?" I asked.

During the spring and summer of 1994, the senator said, a number of parents from some of the surrounding school districts contacted him about their concern over "the mandatory reading of some of Mark Twain's novels." ("How many parents?" I later asked. "About ten," he responded.) They were worried about "the issue of children having to be exposed to the word 'Nigger' Jim—that it was degrading and that it wasn't appropriate for the classroom in these various school districts, and parents were really set back by it and wanted something done about it." But the presence of "the big N-word," as the senator had put it in the press, was not the only problem. There was also the issue of the characterization of Jim: "If you continue to tell people, particularly young, impressionable minds, that they come from a long line of individuals that are quote unquote 'no good,' that have always been in subservient positions, do you just kind of reinforce that and continue to deflate their self-esteem, or should you be about building positive self-esteem?"

I tried thinking about Jim as "no good" and found the idea, in Huck's words, "interesting, but tough." Earlier in the year, the head of the English department at the National Cathedral School had cited the same concern as one of the reasons for taking the novel out of the curriculum. I recalled a response that had run on the *Washington Post* letters-to-the-editor page from Robert T. Fagan of Houston:

> By the twisted standards of modern education, [the characterization of Jim] is an understandable concern. After all, [Jim is] a man whose love for his wife and child persuades him to risk life and limb a second time to liberate them from enslavement, a man whose compassion causes him to routinely stand double watch on the raft so Huck can sleep longer. . . . In short, no matter how mistreated he is by the racist world around him, Jim retains the one thing that "civilized" society cannot take away from him: his own good character.

Perhaps the real problem, I suggested to Senator West, was that Twain's novel was badly taught. Perhaps the students didn't realize that it makes white people look ridiculous or worse, and that Jim is really the hero. "I don't know how to respond to that," he said, adding that he didn't think a literature class was the proper forum to "expose kids to racism."

"What do you think literature can do, ideally?" I asked him. "What's the function of teaching fiction? Of having students read novels?"

"You've lost me on that," said the senator.

"I'm just wondering. . . . Why not just teach history? Why do we bother teaching literature if literature is so complicated and dangerous?"

"It kind of opens your mind up," said the senator. Literature offers "a flavor of how people thought, how people felt about different issues at a particular time in history. It gives you that particular flavor from an individual's perspective. I think that all of us would want to get that. I look at the *Autobiography of Malcolm X*, and it gives me some idea of what he was going through—as opposed to, I guess, just looking at it in some sort of chronology of events that occurred during a particular point in time."

I pointed out that his amendment could have prevented any state funds from being spent on the *Autobiography of Malcolm X*.

The senator was taken aback. "It would not have."

"The word 'nigger' is in that book," I observed.

The Senator was insistent. "No, it would not have." Clearly he believed that the *Autobiography of Malcolm X* would pass the test of "redeeming educational value" and that *Huckleberry Finn* would fail.

The amendment died in the House, but Senator West told his constituents "I tried to do what I could do. And if they wanted to continue the battle, I'd be more than happy to do so next legislative session."

We cordially agreed to disagree. I thanked him for his time and left.

*   *   *

Austin, March 1995. I walked past the Littlefield Fountain on my way from the library to Garrison Hall, where I was to teach my 11 a.m. class. Major George W. Littlefield had fought for the Confederacy at Shiloh, Chickamauga, and other Civil War battles. After the war he made a fortune in land and cattle and was an early benefactor of the University of Texas. He had set up a library fund to buy "anything that contributes to the greater glory of the South," as a colleague had put it soon after I arrived. Without a collection of "historical materials" at a Southern institution of higher learning, Major Littlefield believed, "the protests of patriotic societies against the misrepresentation of the South are 'as sounding brass and tinkling cymbals.' . . . I am anxious to see something done that will begin the foundation for acquiring a history, in which the South may be accorded her just rights." I smiled to myself. Last year I suggested that Littlefield money be used to buy something that genuinely contributed to "the greater glory of the South," by my lights: microfilm of nineteenth-century black Southern newspapers. Research librarian Cheryl Malone agreed, and I had been able to teach a new graduate seminar on the nineteenth-century African-American press centered on this newly acquired material.

I stopped to read the inscription on the fountain, as I often do when I walk by.

> To the Men and Women of the Confederacy who Fought with Valor and Suffered with Fortitude / that States Rights be Maintained and Who Not Dismayed By Defeat Or Discouraged By Misrule. . . .

I always pause there. "Misrule" indeed. The familiar code word for Reconstruction. I found the word as inappropriate as Twain found the inscription on the monument in New Zealand to white men "who fell in defence of law and order against fanaticism and barbarism"—and for the same reason. Only through the lens of a deep-seated racism could Maori patriotism be considered "fanaticism and barbarism," Twain felt, and only through the same lens could the exercise of political power by blacks during Reconstruction be considered "misrule." The word was a capsule summary of the Dunning school of Southern history, dedicated to promulgating the myth that giving blacks political power after the Civil War had been an unmitigated disaster from which the "Redeemers" had had to rescue the South. Historians now recognize that corruption was an equal-opportunity seducer during Reconstruction, distributed fairly evenly among whites and blacks, and that the majority of blacks exercised their new political power responsibly. It took historians several generations to reject the perspectives of the Dunning school as false and to understand the level of betrayal involved in the nation's troubling retreat from racial justice after Reconstruction.

I always look forward to the point in the semester where I can ask my students what the Littlefield Fountain says. They walk by it several times a day, but few of them ever really stop to read it. And those who do rarely give it a second thought. But it is a key point of departure for today's class, our third on *Adventures of Huckleberry Finn*.

"Mark Twain starts writing this novel in 1876," I begin. "What's happening in the country then? And during the next few years, when he puts the book aside before coming back to it?" Some of the English majors and business majors stare at me blankly. An American Studies or history major usually pipes up at this point with a rundown of the election of 1876, the Compromise of 1877, and the breakdown of Reconstruction. At that point I lecture for a few minutes.

I believe it is significant, I tell them, that Twain both began and abandoned this book during the summer of 1876, one hundred years after the signing of the Declaration of Independence. That summer marked the greatest formal celebration of liberty in the nation's history, but it also brought some of the worst racial violence the country had ever seen. A host of particularly brutal racial clashes in July—such as the Hamburg and Ellerton Massacres in South Carolina—were reported fully in the New York papers that Twain read, alongside columns quoting centennial speeches about the excellent health of liberty in America.

I point out that the promise of genuine freedom and participation in the polity seemed within sight for many African Americans during Reconstruction, but that racist opposition to these gains had been building for some time. The tactics of such terrorist groups as the Knights of the White Camellia, the White Brotherhood, the Rifle Clubs of South Carolina, and the Ku Klux Klan included intimidation, physical force, bribery at the polls, arson, and murder, as Twain was well aware. I circulate a list of headlines Twain had clipped from daily newspapers in 1873, including A COURT HOUSE FIRED, AND NEGROES THEREIN SHOT WHILE ESCAPING and KU KLUX MURDERS.

I remind my students that the election of 1876 led to the "withdrawal" of federal troops from the South and marked a key change in national policy. The intimidation that white supremacists had had to carry out at night could now be accomplished in broad daylight. The official government policy was now to look the other way as thousands of African Americans were effectively re-enslaved through such means as sharecropping, lynchings, and the convict-lease system. As Reconstruction collapsed, the hypothetical fear Frederick Douglass had expressed in 1862 actually came to pass: black people had been emancipated from "slavery to individuals, only to become slaves of the community at large."

"What do you think it would have felt like to see all this happening if you were black?" I ask.

"I'd have felt really disappointed, let down," says one student,

"Like the rug had been pulled out from under me," says another.

"How do you feel when you read *Huck Finn* and that exciting journey to freedom somehow ends up with Jim in chains in that ramshackle shed on the Phelps farm in Arkansas?" I ask.

A student smiles: "The same way."

They have spent weeks reading other works by Twain and have learned to detect the complicated ways in which he sometimes manipulates us as readers to make his point. We have discussed the section in *The Innocents Abroad* titled "The Ascent of Vesuvius," where he strings us along with chapter after tedious, interminably digressive chapter—only to briskly dispatch "The Descent of Vesuvius" in one sharp, short half page. "That's what it must have seemed like to him at the time, the trip up compared with the trip down," one student suggested. "He's trying to give us a taste of it as readers." Yes. And we have talked about Twain's puzzling story in that same book of a man who watches a mule fall through his roof time after time after time, only to say, long after the reader's patience is well beyond exhaustion, "This thing is growing monotonous." A student said, "It's like Twain is bored and wants to write about being bored in a way that we'll know he was bored but still stay interested in his book." Yes. They have studied Twain's earlier use of irony in his satires on racism toward the Chinese in San Francisco, and they have read works by black writers that use "signifying" indirection to make hard-hitting, explosive points. (A favorite text is Paul Laurence Dunbar's "An Ante-Bellum Sermon," in which a preacher's sermon on Exodus is used to address the possibility of liberation not only in biblical and antebellum days but in the 1890s, when Dunbar is writing.) So they are with me when I follow up on what the structure of *Huckleberry Finn* may be saying about the period of history in which the book was written.

I suggest our contemporary sense of just how far off freedom was in that summer of 1876 may help explain why Twain decided to smash the raft that was to carry Huck and Jim to freedom. And the confirmation of these fears during the next eight years may well have induced him to end his novel with a farce that reflected the fate of "freedom" for African-Americans in the post-Reconstruction era in both the South and the North.

"So we're supposed to feel sort of let down when it all falls apart and Tom Sawyer turns it into a big game on the Phelps farm?"

"I think we are. What Bernard DeVoto called the 'chilling descent' of the novel's ending mirrors the equally chilling descent embodied in that chapter of history. As Du Bois put it in his book *Black Reconstruction*, 'The slave went free; stood a brief moment in the sun; then moved back again toward slavery.' *Huckleberry Finn* presents itself as a simple boys' book as slyly as the traditional trickster tale presents itself as a simple animal story. Just beneath the surface, however, it dramatizes, as perhaps only a work of art can, both the spectacular boldness of the promise of liberty and justice for all and the nation's spectacular failure to make that promise a reality."

I pass around the McGuinn letter, in which Twain said that "the shame is ours, not theirs, & we should pay for it." We talk about Twain's own sense of guilt about how black people had been treated in America—

about his move from Hannibal to Elmira, about how long it took him to question the norms that surrounded him during his childhood, about the forms that questioning took once it began.

A young man in a plaid shirt raises his hand shyly. "I'm not sure how to say this, but what if we just don't feel bad. What if we kind of enjoy the stupid pranks they play. I mean, it's not that we want Jim to suffer, but we sort of forget about how he feels and just get into watching Tom and Huck plot all that crazy stuff. I mean, are we supposed to enjoy it, sort of find it funny—and then feel guilty that we wanted to find out what happened?"

"Could be," I respond. "Maybe Twain wants us to feel partly disappointed that the quest for freedom gets all smashed up like the raft, and maybe he wants us to feel somewhat complicitous with those who let that happen."

A young man who hasn't said much all term raises his hand. "I think maybe Twain himself thinks it's all pretty complicated. I mean, part of him identifies with Huck, part of him with Tom, part with Jim. That's what makes it a book that's believable."

A girl a few seats away from him nods in agreement. "Twain's pretty sly, when you come to think of it. I mean, he just drops the whole quest for freedom just like that—just like the whole society did—and lets this slave who, as it turns out, *is* free suffer through all that stupid garbage. Just like the South turned back the clock on black people's freedom after the federal troops left. But Twain gets us kind of absorbed by Tom Sawyer's games, and maybe we even forget for a few minutes that it's Jim who's suffering through all this—and at the end we feel like slime."

"Slime? What do you mean?"

"Well, it's like *we* let it happen. It's like we were there and we didn't do anything about it. We didn't get mad and put the book down and yell at Tom to stop or at Twain to shape up—not that it would have done any good. We just sort of went along. Which is exactly what most Americans did when all that garbage was coming down in the 1880s and 1890s."

"I think Sarah's right," says a young man sitting near the window. "We're supposed to forget that there's a human being who is chained and hurt in that shack, and then we're supposed to feel guilty that we forgot. It's like the 'Ascent of Vesuvius' and the 'Descent of Vesuvius.' Twain's trying to give us a way of feeling what it was like to be living through the time when he was writing, even though all he's got to make us feel it is words on a page."

"How do we know Twain hasn't forgotten what Jim's going through?" I ask. "How do we know he isn't just using him as a plot device?"

Sarah raises her hand again. "Jim gives up his freedom to stay with Tom Sawyer. That's a bigger sacrifice than anyone else makes in the entire book. And he gives it up to nurse a real jerk who's put him through all this crap. And then, when he gets back to town, they don't even take his chains off. They don't even give him real food. Huck says he wished they would, but somehow they just forgot. That crowd isn't good enough to lick Jim's boots. *We* see that. We see that because Twain lets us see that."

The bell rings. As I start down the stairs, my thoughts wander to the class that has just ended—and to the articles in the press during the last few weeks about *Huckleberry Finn* being taken out of the curriculum at the National Cathedral School in Washington, D.C., and at West Hills Middle School in New Haven, Connecticut. The story was always the same whether the book was being challenged in Washington or New Haven or Modesto, California; Kinston, North Carolina; Portage, Michigan; Houma, Louisiana; Mesa, Arizona; Plano, Texas; Erie, Pennsylvania; or Tempe, Arizona—all places where *Huckleberry Finn* had been challenged in the 1990s.

I think about the loss to black students when this book is taken out of their classrooms. I recall the response of Bill Matory, a black senior at St. Alban's school in Washington, D.C., to the prospect of that school's taking *Huck Finn* off the reading list: "It's like taking a big part of America's past away from us. As an African-American male, you must understand why the book was written and how it was written. And we are smart enough to understand that." And I think about a black college professor named Jocelyn Chadwick-Joshua. In a suburb of Dallas in the fall of 1993, at a town meeting at which her state representative, her minister, her church, and virtually all of her black friends and neighbors were assembled in support of the other side, she found the courage to take the stage and say:

> It would be a travesty of insurmountable proportions for you to agree to ban this novel when we have such an honorable man of color depicted here. We're telling our children that we still have to deny the fact that we have overcome insurmountable odds. I don't think we want to do that.

In the high school she attended in Texas in the 1970s, the assigned American literature textbook was still the 1958 edition of *Adventures in*

American Literature, She remembers clearly the view of slavery presented in that volume: "Negroes have given America a wealth of song," the book stated, in a section that reprinted the words to several spirituals. It urged students to "let your imagination re-create the scenes that gave rise to the spirituals: men and women picking cotton in the fields; men loading heavy bales on barges, with one rich voice singing out the varying lines and the whole company joining in the refrain." The depiction of slavery in this sanitized, "picturesque" scene, she notes, was "devoid of 'nigger,' hell, poor white trash, lynch mobs, and indifference," as was the textbook used in her history class. Had her "parents and some discerning teachers" not provided her with "supplemental books to read and discuss with them to counteract that depiction," she "would have had a very different impression of my people and Euro-Americans. Among those books was Mark Twain's *Adventures of Huckleberry Finn*."

I think about what black students lose when this book is removed from their classrooms. And I think about what white students lose—and how my own life would have been different if I hadn't been assigned to read it and write about it in high school.

If W. E. B. Du Bois was right that the problem of the twentieth century is the problem of the color line, one would never know it from the average secondary-school syllabus, which often avoids issues of race almost completely. Like a Trojan horse, however, *Huckleberry Finn* can slip into the American literature classroom as a "classic," only to engulf students in heated debates about prejudice and racism, conformity, autonomy, authority, and justice. It puts on the table the very questions the culture so often tries to bury, challenging readers to confront the complex history that shaped it—a history that's often easier, for both blacks and whites, to simply ignore.

Ultimately this is a book for children—for it teaches them what they need to know to make it to adulthood whole. It can teach them to challenge their parents and teachers and to question society's laws. It can make them ask impertinent questions of their history books and knock local heroes off their pedestals. It can encourage them to suspect organized religion and to doubt the truth of most of what they've been taught. Do we really want our children to be this irreverent? I think we do: for Twain teaches them irreverence toward tyranny, injustice, hypocrisy, and fraud—which will come in handy as they struggle to straighten out the mess we've made of our world. (Irreverence, Twain once wrote, "is the champion of liberty & its only sure defence.")

We continue to live, as a nation, in the shadow of racism while being simultaneously committed—on paper—to principles of equality. As Ralph Ellison observed in our 1991 interview, it is this irony at the core of the American experience that Mark Twain forces us to confront head-on in this novel. Irony, history, and racism all painfully intertwine in our past and in our present. And they all come together in *Huckleberry Finn*. Because racism remains endemic to our society, a book like *Huck Finn* can explode like a hand grenade in a literature classroom accustomed to the likes of *Macbeth* or *Great Expectations*,

History as it's taught in the history classroom is often denatured and dry. You can keep your distance from it if you choose. Slaveholders were evil. Injustice was the law of the land. History books teach that. But they don't require you to look the perpetrators of that evil in the eye and find yourself looking at kind, gentle, good-hearted Aunt Sally and Uncle Silas. They don't make you understand that it was not the villains who made the system work but the ordinary folks who did nothing more than fail to question what was going on around them. That lesson is troubling and dangerous and disruptive and crucial.

As I walk to the library, I pass a statue of Jefferson Davis and the marble inscription on Littlefield Fountain. I think about how few demands history places on us in daily life, how easy it is to distance ourselves from a past we've moved beyond, how easy it is to keep statues and stone inscriptions in their places and out of mind, if that's where we want them.

And then I think about *Huckleberry Finn*. What it does to me every time. What it does to my students. And I remember why reading history alone is never enough.

Fiction can drag you in and drag you under. It can grab you by the throat and punch you in the nose, or thrust your nose into foulness so deep the smell is suffocating. When accomplished fiction writers expose the all-too-human betrayals that well-meaning human beings perpetrate in the name of business-as-usual, they disrupt the ordered rationalizations that insulate us from pain.

Novelists, like surgeons, cut straight to the heart. But unlike surgeons, they don't sew up the wound. They leave it open to heal or fester, depending on the septic level of the reader's own environment.

As David Bradley recently observed, "If we'd eradicated the problem of racism in our society, *Huckleberry Finn* would be the easiest book in the world to teach."

*   *   *

A few years ago, after a class like the one I had just finished teaching, I overheard a student ask a classmate as they left the room, "Does all this mean *Huck Finn* can't be the book I loved before I knew all this stuff I know about it now? The romance of the river—the great adventure—is it possible to still read it that way anymore?" They were gone before I heard her friend's answer. But I often think about her question.

I recall that passage in *Life on the Mississippi* in which Twain draws a contrast between the river the passenger sees and the river seen by the steamboat pilot. The passenger, Twain writes, "saw nothing but all manner of pretty pictures in it, painted by the sun and shaded by the clouds, whereas to the trained eye these were not pictures at all, but the grimmest and most dead-earnest of reading-matter."

> . . . when I had mastered the language of this water and had come to know every trifling feature that bordered the great river as familiarly as I knew the letters of the alphabet, I had made a valuable acquisition. But I had lost something, too. I had lost something which could never be restored to me while I lived. All the grace, the beauty, the poetry had gone out of the majestic river! . . . All the value any feature of it had for me now was the amount of usefulness it could furnish toward compassing the safe piloting of a steamboat.

Was this what my student was trying to say? Had I turned the "poetry" of a book she had loved into the "most dead-earnest of reading-matter"? The passage continues, however, as follows:

> I still keep in mind a certain wonderful sunset which I witnessed when steamboating was new to me. A broad expanse of the river was turned to blood; in the middle distance the red hue brightened into gold, through which a solitary log came floating, black and conspicuous; in one place a long, slanting mark lay sparkling upon the water; in another the surface was broken by boiling tumbling rings, that were as many-tinted as an opal; where the ruddy flush was faintest, was a smooth spot that was covered with graceful circles and radiating lines, ever so delicately traced; the shore on our left was densely wooded, and the sombre shadow that fell from this forest was broken in one place by a long, ruffled trail that shone like silver. . . .

Twain hasn't lost the river at all—for he evoked that wonderful sunset *after* he had supposedly "lost" the ability to appreciate it. Twain tells us that he learned to read the river like a book; so now we have learned to read his book like a river. But we, too, may gain more than we lose, just as Twain gained a new river without really losing the old one. We may know more about precisely what makes the surface shimmer so iridescently, or about why certain dangerous shadows need to be watched with great care. But there is still shimmer. And there is still danger.

RIPPLES AND REVERBERATIONS

Was he jester or Jeremiah? Funny man or moral fabulist? Illusion master or delusion blaster? Was he an escapist or an escapee? Poor man's pal or rich man's pet? Boat rocker or captain of the ship?

Twain was all of the above, and in reshaping himself first in one image, then another, he was the embodiment of that most American of traits: the ability to invent—and reinvent—oneself. He preened and paraded. He stretched his fifteen minutes in the limelight to hours, days, years, cultivating his fame with the diligent attentiveness some devote to cultivating African violets. What was he known for? Why, for being Mark Twain, of course. Whatever that meant. And it meant very different things to different people, then as well as now.

Which Twain do we choose to embrace? What if we reject none of them? What would it take to acknowledge the complexity and diversity of this man? Can the laughter he so regularly induces be both cathartic *and* catalytic, an agent of release and reprieve but also of potentially revolutionary changes in consciousness?

When Mark Twain speaks to us across time, are we listening to all the resonances and echoes of that familiar voice or only some? Which ones become salient? Canonical? Iconic? Which ones are buried or denied? "Mark Twain," like "the United States of America," is an idea: words on paper that become what we choose to make of them.

Sightings

Known to Everyone and Liked by All
—slogan on the Mark Twain Cigar

July 1995. As I ride in a Metro-North railcar, a life-sized Mark Twain in a clubby leather armchair with a book on his lap peers down at me genially from an ad for Bass Ale ("What Can Compete with Life's Simple Pleasures?"). A more cynical-looking but just as life-sized cardboard cutout Mark Twain greets me when I enter a local bookstore. I buy Jimmy Buffett's new CD and find a song inspired by Twain's *Following the Equator*. The detective hero of *Black Betty*, the Walter Mosley mystery I am reading, launches unexpectedly into a defense of *Huckleberry Finn* in the middle of the book. My twelve-year-old son, a sci-fi and comedy fan, rents a video with the promising title *Unidentified Flying Oddball*, which turns out to be a space-age remake of *A Connecticut Yankee*. The man is everywhere.

July 1995 was a busy month for Mark Twain. He made personal appearances in Indianapolis, Reno, and Charleston; Kansas City and Fredonia, Kansas; Muscatine, Iowa; Hannibal, Missouri; Elmira, New York; Grant, Nebraska; and Ponca City, Oklahoma. He contributed to obituaries for the typewriter (Smith Corona had just filed for bankruptcy); to a food column on New Orleans and travel columns on Germany and Hawaii; to articles on the Cincinnati city budget, caving, steamboating, language use, current film releases, India, Bosnia, Rush Limbaugh's hometown, an unexpected family reunion in Cairo (Egypt, not Illinois), U.S.–Japan relations, Arthurian legend, and Judy Collins. He held forth on statistics, lawyers, intellectual-property rights, patriotism and the Fourth of July. He sparked a lively debate among inmates at Utah State Prison over the differences between "borrowing" and "stealing," and he induced some two dozen teachers to spend a week in Hartford girding themselves with strategies to deal with what happens when they bring his most famous and most explosive novel into their classrooms.

Clearly, reports of his death have been greatly exaggerated.

Suppose a Fearless Semiotician were to tackle just one of these Twain "sightings." He or she might "read" the Bass Ale ad as follows: We see an impeccably attired icon of democracy who is clearly at home with luxuries like imported ale—indeed, who has come to view them as "Life's Simple Pleasures." Twain's presence is designed to domesticate the foreign, to give red-blooded Americans permission to relish an imported product like Bass Ale without feeling guilty or effete. In ways that are unthreatening and positive, Twain manages to be both broadly democratic and unabashedly elitist. He may well have been a "man of the people" and one of America's most famous chroniclers of the simple pleasures of life, but the image we see here, from the burnished tones of the background to the costly and impractical white suit, is one of an aristocrat at ease. The ad suggests that the ultimate fulfillment of the American Dream may be creating a life that lets you relax in a leather chair with a glass of English ale.

How do we reconcile the urbanity and sophistication projected in the Bass Ale ad with the sheer silliness of spacemen running around King Arthur's court? Our Fearless Semiotician has come down with a sudden migraine and left the room.

Twain's image has been used to sell writing paper, billiard tables, and riverboat travel—plausible given his fondness for all three. But natural gas? The advertising copy strains for what remains at best a rather fugitive connection to Twain. His trademark white suit inspires a dry-cleaning establishment in North Carolina to design an ad featuring a frowning Twain in a soiled white suit holding a turkey-drumstick: "Though Mark Twain was fond of fowl, he liked his suits snow-white. We'd have dry-cleaned those foul stains." As Twain himself said, "Many a small thing has been made large by the right kind of advertising." The most familiar images of him in ads show an avuncular Twain with his trademark shock of white hair (his only competitors in the field of widely recognized white-haired popular icons today are Colonel Sanders, of fried chicken fame, and Albert Einstein).

Businesses recognized the potential of Twain's face to move their products as early as the 1870s, when a company that dealt in "White & Fancy Goods" and sewing machines pictured Twain in their ads, as did a firm of "Plumbers, Steam and Gas Fitters, Tin, Copper and Sheet Iron Workers." After registering his pen name as a trademark with the patent office, Mark Twain allowed it to be used to advertise a new style of collar ("I

WHAT CAN COMPARE WITH LIFE'S SIMPLE PLEASURES?

Samuel Langhorne Clemens (Mark Twain), 1905.

BASS HELPS YOU GET TO THE BOTTOM OF IT ALL.

11. *Advertisement for Bass Ale that appeared on MetroNorth commuter trains in 1995 (Printed with permission of the Guinness Import Company)*

think it is time the name should be connected with something useful—it has been confined to the aesthetic and ornamental long enough") as well as Mark Twain Whiskey, Mark Twain Tobacco, and Mark Twain Cigars.

Louis J. Budd, dean of American Twain scholars, has spent a lifetime tracking the vagaries of Twain's public presence. Budd writes,

> More fluid than a mosaic, Twain's image shimmers like endless MTV. The most diligent observers cannot claim full coverage; what they find depends on how and where they watch. . . . That Twain's image dances out there incessantly is undebatable, but under tight focus it blurs, shifts, and splits like a superameba. No literary geneticist will chart all its chromosomes; no cultural anthropologist will achieve a thick description that convinces most peers; no culture critic can harmonize all the impressions, often emotional and subconscious rather than cerebral.

"The Twain icon," he says, "is a gallery itself of not only images but personalities; intricacy is the simple key here. To a unique degree Twain enshrines different, even differing, values for a span—now a cable-spread—of audiences." In *Our Mark Twain: The Making of His Public Personality*, Budd charts the steady rise of Twain's fame in his own lifetime, revealing the energy and talent with which he shaped his public image. Although Twain lectured mainly for the income, Budd notes that he also valued the grassroots exposure it brought. He consistently courted the press, quickly became a master of the new genre of the interview, and "postured superbly, holding the spotlight and yet not making himself ridiculous or boring."

What Twain himself began, the rest of the world has continued. Mark Twain's unfailingly poor judgment in the investing field evidently failed to trouble the founders of the Mark Twain Investment Company in Kansas City, Missouri. And his inability to hold on to whatever money he made doesn't seem to have bothered the Mark Twain Bank of St. Louis. The fact that Twain's one real estate holding—the famous "Tennessee Land" his father bought—was close to worthless when his brother sold it didn't deter Mark Twain Real Estate of Los Angeles (if pressed, they might point out that Orion sold too soon; the land is worth a fortune today). Did the proprietors of Mark Twain's Pizza Land in Metairie, Louisiana, or Mark Twain's Pizza Landing in New Orleans hope that folks might remember that Twain liked the food he ate in that state? (He found the pompano heavenly, but his views on pizza are unknown.)

S.H. CLEMMENS. (MARK TWAIN!)

HARRIS & MOXLEY,

DEALERS IN

White & Fancy Goods,

SEWING MACHINES, &C.

12 Main Street, New London, Conn.

J. Emerson Harris,   Francis G. Moxley.

12. Advertisement using Twain's image to sell "White & Fancy Goods, Sewing Machines, &c." from the 1870s (Photo courtesy of the Mark Twain House, Hartford, Connecticut)

Surely any of the many barbers who struggled in vain to tame Twain's famously untamable mop would have been amused at the notion of a Mark Twain Hair Design in New Orleans.

Despite the fact that in his most famous novel "Mark Twain showed a sovereign contempt for the classroom," as one journalist recently put it, there are preschools, elementary schools, and secondary schools named after him in California, Colorado, Connecticut, Illinois, Kansas, Maryland, Michigan, Missouri, New Mexico, New York, Oklahoma, Oregon, South Dakota, Texas, Virginia, and Washington. There is no college named after Twain, however. Efforts to change the name of the institution a few blocks from the Mark Twain House in Hartford to "Mark Twain Community College" foundered on objections to Twain's use of racially charged language in *Huckleberry Finn*.

* * *

Mark Twain is cited in publications ranging from *Nutrition Today* to *American Banker*. He is claimed as a forebear by groups that include self-made men, pessimists, humorists, humanists, cat lovers, advocates of self-publishing, and enthusiasts of homeopathy. He is summoned as an expert on everything from gardening to grammar, from baseball to business strategies. The affection with which he is invoked in such a broad range of venues would have come as no surprise to Twain, who in a 1908 speech was able to observe, "Mr. Edison wrote: 'An American loves his family. If he has any love left over for some other person, he generally selects Mark Twain.' "

"If cauliflower, as Mark Twain wrote, 'is nothing but cabbage with a college education,' then calabrese and broccoli have the Ph.D.s of the family," observes the gardening columnist in the *Independent*. Each constituency seems to have its favorite Mark Twain quote. The business community, in commentaries on investing strategies, favors "put all your eggs in one basket, and watch that basket." The foreign-policy community, in explaining why conditions in a particular part of the globe are less dire than meets the eye, prefers to paraphrase a comment commonly attributed to Twain: "Wagner's music is better than it sounds." (The remark actually comes from humorist Bill Nye, whom Twain quoted in his autobiography.) Congress-bashers sometimes rely on Twain to do their dirty work, as in "Fleas can be taught nearly anything that a Congressman can" or "Readers, suppose you were an idiot. And suppose you were a member of Congress. But I repeat myself." "Get your facts first and then distort them as much as you please" is almost de rigueur when a spokesman for some concern or other wants to challenge a competing analysis of the situation.

By far the most ubiquitous Twain quote is one he wrote in a note in London on May 31, 1897, to rectify a mix-up. His cousin James Ross Clemens had fallen ill, and newspapers had somehow confused him with Mark Twain, circulating the story that Twain was dying. Frank Marshall White, an American journalist then living in London, who had seen Twain a day or two before, sent a reporter to Chelsea to check on his health. Twain sent the reporter back with a note:

> James Ross Clemens, a cousin of mine, was seriously ill two or three weeks ago in London, but is well now.
>
> The report of my illness grew out of his illness; the report of my death was an exaggeration. MARK TWAIN.

Later that day White called on Twain personally. "Of course I'm dying," Twain said, "but I'm not dying any faster than anybody else." After his visit, White tells us,

> I sent a despatch about Mark Twain's financial and physical condition to my paper that night, in which I embodied in his own words what he had written about his cousin in the morning. The operator who cabled it left out all punctuation marks, as is usual unless they are specifically marked for transmission. The copy editor in New York, preparing the despatch for the printer, began a paragraph with the last clause of the second and last paragraph of the note Mark Twain had written in bed, "The report of my death is an exaggeration," and this by process of repetition became, "The reports of my death are grossly exaggerated."

"To paraphrase Mark Twain, stories about the death of newspapers have been greatly exaggerated," said Roger Fidler of Knight-Ridder Newspapers on the TV series *Science and Technology Week. ID: The Voice of Foodservice Distribution* informs readers that "despite the disappearance of some of the old familiar favorites, there are still plenty of fish in the sea and on the farm. To paraphrase Mark Twain, the reports of seafood's death are greatly exaggerated." Reports of the death of women's tennis, silicon semiconductors, the mainframe computer, the space shuttle, videoconferencing, the satellite-dish industry, the political machine, the real estate market, stenography, the department store, electric utilities, the cassette deck, the peace process in Northern Ireland, British pubs, brand-name colas, inflation, measles, books, defense spending, traveler's checks,

Yiddish, and U.S. sprinting have all been—to paraphrase Mark Twain—"greatly exaggerated," as journalists duly note. So omnipresent is this rhetorical construction that the *Times* of London has named a special category of story after it: the Mark Twain obituary class.

Twain's eminently quotable quotes are pithy and smart. But plenty of people have made plenty of pithy and smart remarks that aren't repeated as often as Twain's. Rather, these quotations have endured by taking on the luster of the personality that produced them. You can trust Mark Twain to be someone who takes you into his confidence and reassures you that he's on your side; someone who can be brutally honest but is never malevolent; someone whose abiding affection for you tempers his irritation; someone who conveys that no matter how foolish your behavior, he will never actually abandon you. Twain's avuncular stance helps take the sting out of barbs that would wound more sharply from less friendly lips; his humor softens his satire and makes it palatable. As a result, he can say things that from anyone else might lead to blows. He is simply allowed to be more irreverent than most people because there is a deep sense (to paraphrase Bill Nye) that he's not as bad as he sounds.

Who but Twain could get away with "it was not that Adam ate the apple for the apple's sake, but because it was forbidden. It would have been better for us—oh infinitely better for us—if the serpent had been forbidden," or "patriotism is usually the refuge of the scoundrel. He is the man who talks the loudest," or "Few things are harder to put up with than the annoyance of a good example." In all of these quotes, Twain conveys a truth that polite society customarily denies: that we all wish it were easier to be good, or that patriotism often covers base deeds, or that exposure to virtue can be more irritating than inspiring since it underlines our own shortcomings. Summarized in this manner, these comments fall flat with a dull moralistic thud. They don't sound that way when Twain says them. "To get the right word in the right place is a rare achievement," Twain wrote. "To condense the diffused light of a page of thought into the luminous flash of a single sentence is worthy to rank as a prize composition just by itself." Or, as he put it elsewhere, "A powerful agent is the right word: it lights the reader's way and makes it plain."

Mark Twain was not the only American writer, of course, to come up with a large number of memorable witticisms. Ben Franklin cornered the maxim market in his day (indeed, *Poor Richard's Almanack* paved the way for Pudd'nhead Wilson's Calendar). Emerson was no slouch at the aphorism game. And Ambrose Bierce delivered himself of many sharp, well-crafted insights into the human condition. But Twain beats all contenders when it comes to being quoted today. He is certainly not sharper than Bierce or Emerson or more down-to-earth than Franklin, but he does have more heart. ("What a bare, glittering ice-berg is mere intellectual greatness," Twain once wrote.) We can't really trust Franklin to care about us, since he seems more preoccupied with what we think of *him*, with being one of those annoying "good examples." We are suspicious of Emerson's aloofness, perhaps, and Bierce's bitterness gives us pause. But we know that we can trust Twain.

Mark Twain is almost as famous for the things he didn't say as he is for the witticisms whose pedigrees are unchallenged. Every year the editors of the Mark Twain Project are hounded by people who want to know exactly where Twain said, "The coldest winter I ever spent was a summer in San Francisco," or "Giving up smoking is easy. I've done it hundreds of times," or "When I feel the urge to exercise I go lie down until it passes away." The answer to all of these questions is: nowhere. Twain didn't say it. Or at least no one has been able to document that he did. Nonetheless these comments are attributed to Twain in a variety of publications every year. Twain gave so many interviews to so many journalists, and had so many conversations recorded by so many acquaintances, that the jury is always out, so to speak, on these apocryphal attributions. But for the moment, at least, we can't link them to Twain.

Every Twain scholar has had this experience: A friend calls. "Please tell me where Twain said such-and-such and I'll be eternally in your debt." You look through all the likely volumes. You run database searches. You query all your Twainiac friends. No quote. I was once offered a case of scotch if I could find the citation for something Twain allegedly said about good whiskey: "Too much is not enough." I haven't found it yet. Robert Hirst, general editor of the Mark Twain Project, believes that Twain leads Benjamin Franklin and Abraham Lincoln in misattributed quotes. Here are some things he didn't say that the editors at the Mark Twain Project are often asked to run down:

When the end of the world comes, I want to be in Cincinnati because it's always twenty years behind the times.

There is nothing so annoying as to have two people go right on talking when you're interrupting.

When I was a boy of fourteen, my father was so ignorant I could hardly stand to have the old man around. But when I got to be twenty one, I was astonished at how much the old man had learned in seven years.

Two of the most famous quips widely attributed to Twain—"Everybody talks about the weather, but nobody does anything about it" and "Politics makes strange bedfellows"—were actually the work of his friend, neighbor, and literary collaborator Charles Dudley Warner.

* * *

Like his face and his bons mots, Mark Twain's characters have acquired symbolic lives of their own in the iconography of American popular culture. In October 1995 photographer Jim Goldberg introduced his documentary exhibit of pictures of teenage runaways at the Corcoran Gallery of Art in Washington, D.C., with the words "Huckleberry Finn is dead. It's not that kind of life anymore." That same month Nancy Matthews, spokesperson for a community-service center for homeless and runaway youths in Fort Lauderdale, told a reporter, "It's no longer like Huckleberry Finn. You can end up dead or in deep trouble with drugs and gangs." Huck's father's drug of choice may have been moonshine, and the gangs of Huck's world may have had names like the "Shepherdsons" and the "Grangerfords," but is this abused runaway child all that different from the teenage runaways in Fort Lauderdale or those in the pictures at the Corcoran? Making Huck a poster child for the freedom and innocence of a bygone era is a mistake: experiences like being locked in a cabin with a drunk who happens to be your father pointing a loaded shotgun at you (during a delirium tremens attack) or watching as your friend Buck is senselessly murdered only a few feet away from you age a body fast. Huck, in fact, lives in a world of nauseating violence. Erase the danger and terror at the heart of his novel by filtering it through a wash of nostalgia, and you're left with a carefree childhood pastoral. That is not the book that Mark Twain wrote.

Similarly, erase the selfishness, deceit, and manipulation in the character of Tom—in both *Tom Sawyer* and *Huckleberry Finn*—and you're left with a harmless perpetrator of high-spirited hijinks. Twain's Tom is full of youthful energy, to be sure, but his character is more complicated than that. Consider, for example, his callous complicity in the re-enslavement of a free Jim in the last portion of *Huckleberry Finn*. Or consider the whitewashing of the fence. The fence-painting contest is a central event in Hannibal's Tom Sawyer Days each July. What exactly is fence painting all about? For the contestants in Tom Sawyer Days it is a speed-painting event in a smorgasbord of competitions leading up to the selection of a seventh-grader who will represent the town as "Tom Sawyer" during the coming year. But for Tom Sawyer himself the challenge was not how fast he could paint the fence but how quickly he could finagle his way out of doing any painting at all. For the would-be Tom Sawyers, whitewashing a rectangle of board is also supposed to be fun. After all, that's how Tom Sawyer "redefined" his assigned work that famous sunny afternoon. Good, clean fun. Well, yes. And no. The scene sticks in people's minds not because it introduced them to a marvelous but neglected form of entertainment (fence painting doesn't seem to have caught on as a leisure activity in Hannibal, or anywhere else, for that matter) but because they relish Tom's ingenious plan to get out of work—and get rich while doing it. What from one perspective may look like old-fashioned "good, clean fun" is actually a parable of winning through manipulation, lining one's pockets by persuading folks that they're really having a grand old time worth paying for—which, as it turns out, is what's involved in the Hannibal tourist industry's efforts to lure visitors there to watch boys whitewash a fence.

What fence-painting competitions are to Hannibal, Missouri, jumping-frog contests are to Calaveras County, California, where three to four thousand frogs with names like Rosie the Ribiter and Heavy Metal compete in front of some thirty thousand tourists, setting world records and winning purses of fifteen hundred dollars. Truth be told, "The Celebrated Jumping Frog of Calaveras County" is a story about a man with a propensity for betting who is thwarted by a cheat. Take out the cheating part and what's left is—good, clean fun? That's what the participants say. The animal-rights advocates who periodically protest such contests disagree. But keep in mind that none of this has much to do with anything Mark Twain wrote—precisely the point his daughter Clara made in 1949 when she sued Columbia Pictures for three hundred thousand dollars for having "deformed and mutilated" her father's work and damaged his literary reputation in the 1948 film *Best Man Wins*. She complained that the studio had turned "The Celebrated Jumping Frog of Calaveras County" into a "corny love story," that the film "did use the name of Jim Smiley, the chief character in her father's tale, and made some reference to 'Dan'l Webster,' the frog," but that there all resemblances ended. Her suit demanded that Columbia delete Twain's name from the credits and from promotional materials.

Twain's works are constantly being remade and reinvented on stage, screen, and television, and in a variety of musical forms. Indeed, on this front he may well have outdistanced any other author in the history of the world, with the exception of Shakespeare. Sometimes—as in *Best Man Wins*—little more than the characters' names are the same; on other occasions versions of Twain's work in media other than print evoke the spirit of the original in ways that Twain himself might have admired.

Huck Finn's debut in a popular song may have been the widely recorded "Huckleberry Finn," a 1917 ragtime tune by Joe Young with words by Sam M. Lewis and Cliff Hess. (Young and Lewis were the team that gave us "Five Foot Two, Eyes of Blue" and "How Ya Gonna Keep 'Em Down on the Farm?"):

> CHORUS: Huckleberry Finn, If I were Huckleberry Finn; I'd do the things he did, I'd be a kid again. You'd always find me out fishin', beside a shady pool; Wishin' there never was a school. . . .

(The same popular image seems to have been operative in the early 1960s in the admissions office of Brown University, which informally referred to applicants who had potential but who would rather be floating down a river on a raft as "Huck Finns.")

A more nuanced version of Huck might have emerged in the musical on which the great German-Jewish composer Kurt Weill was collaborating with the American playwright Maxwell Anderson when Weill died of a heart attack just after turning fifty; the score to *Raft on the River* remained unfinished at Weill's death. Operas, oratorios, orchestral suites, and choral works based on "Jumping Frog," "The War Prayer," *Roughing It, A Connecticut Yankee,* "The Awful German Language," and "The Golden Arm" have made their way onstage in recent years in Boston, New York, Nevada, and elsewhere. Singer-songwriter Jimmy Buffett's mellow country, folk, and calypso-style recordings are rife with references to Twain's work: the raft and riverboat images in Buffett's "Barefoot Children" recall *Huckleberry Finn,* and "The Remittance Man" and "That's What Living Means to Me" were inspired in part by *Following the Equator.*

Stage productions of many of Mark Twain's works appeared shortly after their initial publication, and film adaptatations have appeared at a fairly regular clip since 1909, when Thomas Edison produced an abbreviated *Prince and the Pauper.* The 1990s alone saw new versions of *The Prince and the Pauper,* "The £1,000,000 Bank-Note," *A Connecticut Yankee in King Arthur's Court, Tom Sawyer,* and *Huckleberry Finn*—often updated to reflect contemporary developments and concerns.

The Prince and the Pauper remains one of Twain's most frequently dramatized works. In the 1880s Twain himself played Miles Hendon in

13. *Sheet music for the widely recorded 1917 song, "Huckleberry Finn."*

a version of the book for family and friends (with a script by his wife and his daughters playing the title roles). He attended a performance of the story staged by a group of immigrant children on the Lower East Side in 1907. In 1915 the first feature-length film adaptation, as Kent Rasmussen notes, made early use of the split-screen technique in its attempts to do justice to Twain's story; a later film features identical twins as Tom and the prince. But how do you remake this novel in an era when traditional royalty has lost its luster? Who would be the equivalent of a prince for young people in the 1990s? A rock star, of course—or so the producers of the 1995 made-for-TV movie *Prince for a Day* wagered when they cast actor Joey Lawrence as both an obnoxious rock idol and the pizza deliveryman to whom he offers ten thousand dollars to impersonate him. The verdict of one reviewer: "It's all very good-natured silliness, but silliness it is. . . . At least poor Mark Twain has the excuse of being dead." (The more mediocre the remake, the more good reviews Twain's originals garner in the press.) Truer to Twain's book is the Mickey Mouse cartoon version, which appeared in 1990.

"The £1,000,000 Bank-Note," is another longtime favorite on the remake circuit. The first sound adaptation, called *Man with a Million*, starred Gregory Peck and was released in 1954. The most recent version, *A Million to Juan* (1994), gives Twain's story a Latino twist. Actor-director Paul Rodriguez plays Juan, an appealing down-and-out widower with a young child, struggling to survive in east L.A. While Juan is selling bags of oranges on the street corner, a white limousine appears and a stranger hands him a check for one million dollars, with the condition that he return it (in exchange for a gift) in a month. This lively and engaging film works on its own, and although Rodriguez of necessity takes a number of liberties with the letter of Twain's story, he remains remarkably faithful to its spirit.

The Prince and the Pauper and the "The £1,000,000 Bank-Note" are likely to continue to attract producers and directors as long as the fantasy of transforming oneself instantaneously into one's polar opposite retains its appeal. The palace may become a rock star's lair and paupers may deliver pizza or sell oranges, but the basic idea remains both compelling and intact. The wealthy in Twain's stories do not necessarily deserve to be rich, nor are they particularly "good." The destitute do not necessarily deserve to be poor, nor are they particularly "bad." In Twain's view, decency, fairness, and compassion are equal-opportunity virtues as likely to be found in the slums as in the palace. In these stories it is chance, above all, that shapes one's material fortunes—and the visual quality of those fortunes (lush palaces versus decaying slums, a tramp's rags versus elegant haberdashery) makes them particularly suited to film.

Filmmakers have also been perennially attracted to *A Connecticut Yankee*. In handling the time-travel plot, twentieth-century directors have tended to favor the latest technology of their own day over the level of technological development reflected in the 1889 novel. In a 1921 version

14. *Juan Lopez and his brothers discuss how to spend the million dollar mystery check in "A Million to Juan," a film inspired by Mark Twain's story "The £1,000,000 Bank-Note" directed by Paul Rodriguez. Pictured from left to right are Bert Rosario, Paul Rodriguez, and Tony Plana (© 1993 Contemporary Twain, Inc. Photo courtesy of Prism Pictures)*

produced by Twentieth-Century Fox, for example, motorcycles were central to the action. In the 1931 sound version, Will Rogers as the Yankee wows the crowd with his automatic cigarette lighter and the parade of the late-model sedans that race to his rescue. Space-age rocketry and robotics feature in the 1979 Disney version titled *Unidentified Flying Oddball*. In the 1989 television version, the Boss—here a twelve-year-old girl played by Keshia Knight Pulliam (of *Cosby Show* fame)—stuns Camelot into submission with an instant camera and a tape recorder: the Boss claims to be able to destroy Merlin by tearing his picture in half, and stops a charging knight dead in his tracks by playing his own threats back to him in his own voice. And in the 1995 Disney feature film *A Kid in King Arthur's Court*, the fourteen-year-old hero shouts "Rock!" when asked to name his weapon in a duel—and brings his foes to their knees with an ear-shattering blast from his portable CD player.

The 1990s have also seen a range of productions of *Tom Sawyer*. The most unusual is probably the 1995 "Wishbone" television version, part of a series broadcast on Nickelodeon and PBS. Here, using the running-monologue/voice-over technique popularized in such television shows as

My So-Called Life, the story is "narrated" by "Wishbone," a Jack Russell terrier whose real-life name is Soccer and who beat a hundred other dogs in the audition for the role of "Tom" (nothing to the twenty-five thousand "Tom" hopefuls in the talent search for David O. Selznick's 1938 film). The most recent film version of *Tom Sawyer* is Disney's 1995 *Tom and Huck*, in which the cinematography emphasizes the spooky, violent, ominous, and menacing dimensions of the world Twain paints in the novel. The Huck portrayed here by Brad Renfro is street-smart and cocky, but also vulnerable in a very contemporary way: a battered child of an alcoholic who interposes a punk shell of feigned indifference between himself and his pain.

The major film version of *Huckleberry Finn* to debut in the 1990s was director Stephen Sommers' 1993 Disney production *The Adventures of Huck Finn*. To his credit, Sommers recognized that many of the previous film versions of the novel perpetuated racist stereotypes. Determined to avoid these, as well as equally offensive racist language, Sommers strategically revised the story he brought to the screen. He succeeded in steering clear of racist stereotypes and language to some extent, but in the process he managed to completely eviscerate the social impact of the novel. If one is not willing to show racists, how can one effectively satirize racism? Twain knew that; he also knew the consequences. For that reason (and others) his novel has been at the center of a firestorm of controversy for over a century. Stephen Sommers' innocuous version of the book may be good Hollywood, but it is bad Twain.

Which stage or screen *Huckleberry Finn* is the best? Each of the many versions produced since the first film adaptation (a 1920 silent movie) has its flaws. Perhaps the version that stays closest to the spirit of Twain's novel is *Big River*, with music and lyrics by Roger Miller, which opened in Cambridge, Massachusetts, in 1984 and moved to Broadway in 1985, garnering seven Tony Awards, including "Best Musical." Roger Miller's score interweaves black and white musical traditions as artfully as Twain blended black and white voices and rhetorical traditions. His stunning contrapuntal montage of bluegrass and spirituals, country and gospel, fugues and rounds, and calls and responses, succeeds to a large extent in capturing Twain's alchemy. Although it has closed on Broadway, *Big River* continues to play in regional theaters in the 1990s. As the interconnectedness of "black" and "white" traditions in all the arts starts to receive the attention it deserves, the daring beauty of Miller's vision—and Twain's, which inspired it—seems all the more contemporary and important.

Stage and screen versions of Twain's works also appear outside the United States, helping to sustain the worldwide popularity—unprecedented for an American author—that Twain achieved in his lifetime. (As early as 1878, Thomas Wentworth Higginson remarked, after having breakfast with two German students in Coblenz, "As for Mark Twain, they all quote him before they have spoken with you fifteen minutes and always give him a much higher place in literature than we do. I don't think any English prose writer is so universally read.") *Big River* played to enthusiastic houses in Tokyo and Osaka in 1985, a stage version of *The Prince and the Pauper* is performed regularly for young audiences by a Japanese repertory company, and an animated series based on *Huckleberry Finn* won some of the highest ratings in Japanese television history when it first aired in 1976 (it has been seen in reruns throughout the world ever since). Highly successful productions of *The Diaries of Adam and Eve* were mounted in the 1980s and 1990s in theaters in Buenos Aires, and film versions of *Huckleberry Finn*, *Tom Sawyer*, and *A Connecticut Yankee* appear regularly on Argentine cable television. Films based on Twain's works run with Hebrew subtitles on Israeli television. *Raja aur Runk*, an Indian film version of *The Prince and the Pauper*, was released in 1968, a Chinese film adaptation was made in 1966, and versions were also produced in Austria (1920), the Soviet Union (1943), Ireland (1969), and Australia (1970). A film version of "The £1,000,000 Bank-Note" is screened annually at Beijing Foreign Studies University, students there have performed scenes from the story, and members of the English faculty have staged "The Man That Corrupted Hadleyburg."

Twain's popularity outside the United States can often be traced, in part, to a native writer who championed his works. For example, Twain was first translated into Chinese at the initiative of Lu Xun, often considered China's greatest writer. He became entranced by the illustrations in a copy of *Eve's Diary* left behind by some Western neighbors and arranged to have a colleague translate the book into Chinese. It was published in 1931 with an introduction by Lu Xun writing under the name Tang-Fengyu. Twain has had a strong following in China ever since. In Brazil Monteiro Lobato, a pioneer publisher and translator whose own work displays the kind of humor we associate with Twain, was probably the first to translate Twain into Portuguese. Many Japanese were introduced to Twain via the numerous translations of respected humorist and scholar Kuni Sasaki, who also adapted *Tom Sawyer* and *Huckleberry Finn* into easy-to-read Japanese versions for children.

A "self-appointed Ambassador at Large of the U.S. of America—without a salary" (in his own words), Twain has indelibly shaped the image of the United States abroad. Teachers in the Netherlands, England, Japan, and Portugal report that young people in those densely populated and highly stratified countries are endlessly fascinated by the unplanned, unmapped physical and social spaces they encounter in Twain's work. Dutch teenagers are "riveted by the idea of going off by yourself to a place where *no one else is*." Readers for whom "planned landscapes . . . both physical and social" are the norm find great appeal in the physical vastness of the world Twain presents, as well as in the social and geographical mobility of its inhabitants. Readers "who tend to feel cramped under overly rigorous school management" are charmed by the relative freedom Twain's young characters enjoy. A professor in Beijing notes that the values projected by Twain's works often come to be regarded as "typically American." "An eager desire to 'strike it rich' and a strong belief in freedom, democracy, equality, and social and racial justice, . . . great sympathy for the oppressed and exploited, [and a willingness to criticize] what he believed to be wrong with the society and the U.S. government" make Twain the quintessential American author for many Chinese.

Mark Twain may be the consummate Rorschach test for anyone who sets out to understand the United States. Like abstract ink blots that elicit highly personal interpretations more revealing of the viewer's psyche than of the nature of the blots, Twain's works help readers around the globe articulate their own concerns and perspectives while ostensibly responding to the issue of what "America" is to them. Peter Stoneley's students at Queen's University in Belfast, for example, are mainly "Irish Catholics living in a British Protestant province, or Protestant Unionists who are British but not on the 'mainland,' and who are worried about being 'sold down the river' by the British in a 'peace process.' " One reason they relish lively debates over the portrayal of Jim in *Huckleberry Finn*, Professor Stoneley suggests, is that discussing representations of African Americans may "help them think about being a marginalized other"—a subject too dangerous to discuss openly in contexts closer to home. In a similar vein, Professor Makoto Nagawara of Kyoto suggests that Twain encourages students outside the United States to think critically about their own cultures and fosters a "wholesome scepticism that refuses to take anything for granted."

Twain's political attitudes continue to shape many responses to his work around the world. Professor Prafulla Kar of the Maharaja Sayajirao University of Baroda believes that Twain's "anti-imperialistic stance," which stands in sharp contrast to "Kipling's strong imperialism," helps account for Twain's enormous popularity in India. His large and appreciative readership in China also owes something to his antiimperialism, along with his exposés of the oppression of Chinese immigrant workers in the United States.

With graceful agility Mark Twain walked a tightrope of tolerable irreverence across a chasm of unforgivable disrespect. He knew just how far he could go in his criticisms of English political institutions, for example, without alienating the affection of English readers. As Peter Stoneley has observed, "Twain challenged specifically English traditions in important ways and also at a trivial level," engaging in clearly improper behavior such as smoking cigars at an inappropriate time of day or appearing on the street in his bathrobe. "On the other hand, he also affirmed our cultural prestige—accepting an Oxford degree, attending a Windsor garden party impeccably attired, telling his best stories at the best London clubs . . . and simply by living here from time to time. So he teased us, but when it mattered, we could trust him not to offend us." Twain maintained the same delicate balance with respect to Germany, a country where he is regarded fondly despite his famously blunt and caustic satire on its language. ("The Awful German Language" has even become an underground classic wherever German is taught, courtesy of teachers who by definition find the language less mystifying than Twain did.) By the same token, the fun Twain had at the expense of Italian guides in *The Innocents Abroad* did not deter Italian critic Mario Materassi from writing an appreciative review of the book on the occasion of a new Italian translation and Twain's many nasty comments about the French, against whom he professed to be hopelessly prejudiced, did not discourage French critic Roger Asselineau from producing a landmark book in 1954 titled *The Literary Reputation of Mark Twain*. "Praise is well, compliment is well, but affection—that is the last and final and most precious reward that any man can win," Twain wrote. Mark Twain won that reward, in his own time and ours. He indeed seems to be "Known to Everyone and Liked by All."

*   *   *

October 1995. A plumber from Clarke Kent ("Out of This World Service") Plumbing is sawing through the wall in the next room to replace a broken shower. I know what he's thinking: What adult in her right mind would sit at her kitchen table, staring intently at her TV, and take nonstop notes on her PowerBook on Road Runner cartoons? I'm afraid

I only make it worse by muttering something about "research." The plumber tiptoes behind me when he has to return to his truck for more tools. I keep taking notes:

"Guided Muscle." A scrawny Coyote carefully seasons his stewpot, first grinding some pepper from a wooden peppermill, then adding a soupçon of oil from a glass cruet. He regards his work in happy anticipation; a table set with a white cloth and eating utensils awaits him at one side. He then extracts the fruit of his culinary exertions from the pot and puts it on his plate. But when he starts to try to use his knife and fork, he is jerked out of his dreamworld into reality: all he has cooked is an empty tin can. As he sits, despondent and forlorn and *hungry*, Road Runner dashes by. Coyote immediately smacks his lips, imagining how delicious the bird would taste, and races off after him.

"Lady, the shower's fixed," the plumber says some time later. "We'll send you the bill." He shakes his head in the driveway when he thinks I'm not looking.

I pick up *Chuck Amuck*, the autobiography of Chuck Jones, the "legendary animation director" who was honored by the Academy of Motion Picture Arts and Sciences in 1996 with a special award for lifetime contributions to the cinema. On page thirty-four Wile E. Coyote chases after his famous prey, Road Runner, with a knife and fork in his hand and a determined glint in his eye. The artist's caption: "The Coyote—Mark Twain discovered him first." In the upper-left-hand margin is none other than Mark Twain himself. Wearing a halo. The caption: "Mark my words, this is my dearest friend."

"One fateful day our family moved into a rented house, furnished with a complete set of Mark Twain, and my life changed forever," Jones writes in this lively volume bursting with sketches of Road Runner, Wile E. Coyote, Bugs Bunny, Daffy Duck, and a host of other Warner Brothers figures he created or shaped in major ways. (Other animation pioneers were entranced by Mark Twain as well, although they didn't leave the paper trail that Jones did. Walt Disney, for example, whose hometown of Marceline, Missouri, was down the road a stretch from Hannibal, devoured all of Twain's works as a child and retained a permanent affection for him. (When plans for Disneyland were being drawn up, "Tom Sawyer's Island"—circled to this day by the steamboat *Mark Twain*— was the only part of the park that Disney designed entirely by himself.) What particularly impressed Jones was Twain's ability to portray quirky,

15. *Sketch of Mark Twain with a halo drawn by the celebrated animation director Chuck Jones (creator of Road Runner, Wile E. Coyote, and other mainstays of Saturday morning cartoons) who credits Twain with being the key inspiration for his art (© 1996 Chuck Jones Ent., printed with permission)*

funny, and yet totally plausible *character* in animals—particularly one memorable coyote he encountered while devouring Mark Twain's *Roughing It* at the age of seven: "I had heard of the coyote only in passing references from adults. [I] thought of it—if I thought of it at all—as a sort of dissolute collie. As it turned out, that's just about what a coyote is, and no one saw it more clearly than Mark Twain." In *Roughing It* Twain wrote:

The coyote is a long, slim, sick and sorry-looking skeleton, with a gray wolf-skin stretched over it, a tolerably bushy tail that forever sags down with a despairing expression of forsakenness and misery, a furtive and evil eye, and a long, sharp face, with slightly lifted lip and exposed teeth. He has a general slinking expression all over. The coyote is a living, breathing allegory of Want. He is *always* hungry. He is always poor, out of luck and friendless . . . even the fleas would desert him for a velocipede. . . . He does not mind going a hundred miles to breakfast, and a hundred and fifty to dinner, because he is sure to have three or four days between meals, and he can just as well be traveling and looking at the scenery as lying around doing nothing and adding to the burdens of his parents.

In Mark Twain's coyote we see the contours of Wile E. Coyote. Whenever Road Runner races past, Coyote imagines him on a platter, steaming hot, evenly browned, garnished with vegetables and potatoes. The madcap chase is always fueled by a hunger as vast as the craggy chasms of the desert landscape in which the action takes place—a chase Coyote is always doomed to lose, his abject humiliation immediately giving way to new plots to trap his opponent. Road Runner and Coyote cartoons, Jones notes, "are known and accepted throughout the world. . . . If you want to laugh, you can do so at any time, whether in Danish, French, Japanese, Urdu, Navajo, Eskimo, Portuguese, or Hindi. 'Beep-Beep!' is the Esperanto of comedy."

Although Road Runner is clearly a bird, he takes his character and abilities from several of Mark Twain's nonfeathered creations in *Roughing It,* among them the jackrabbit. Indeed, Chuck Jones tells us that Twain's chapter on jackrabbits in *Roughing It* "gave me the clue to the speed of the Road Runner." When he perceives himself being shot at, Twain's rabbit "straightens himself out like a yard-stick every spring he makes, and scatters miles behind him with an easy indifference that is enchanting." As in so many cartoons, the rabbit's exit is both dramatic and classic: "He dropped his ears, set up his tail, and left for San Francisco at a speed which can only be described as a flash and a vanish! Long after he was out of sight we could hear him whiz."

The dictionary definition of the word "animation," Jones reminds us, is "to invoke life." He recounts the experience of a young man who came to work at Warner Brothers

shortly after World War II and promptly and proudly wrote home to his grandmother that he was writing scripts for Bugs Bunny.

"I can't understand why you're writing scripts for Bugs Bunny," the old lady responded with some asperity. "He's funny enough just as he is."

Implicit in her comment, Jones says, was the idea that the animators' job

was not to invent what Bugs Bunny did but to *report* his doings. Just as I, at seven, upon reading *Tom Sawyer*, would have been outraged at the suggestion that Mark Twain or anybody else *invented* Tom and Huckleberry Finn and their company. Tom Sawyer happened. He was not imaginary. He was *real*. He still is real. What else can he be but real?

As a child Jones had had a memorable cat named Johnson who had an uncontrollable passion for grapefruit. Johnson would "leave a Bismarck herring, a stick of catnip, or a decayed seagull for a single wedge of grapefruit. For a whole grapefruit, he would have committed fraud or practiced usury." Looking back on his career, Jones remarked: "What grapefruit was to Johnson the cat, Mark Twain became to me." Jones came to understand that "Mark Twain used words the way the graphic artist uses line control." Through his influence on Jones and others, Twain would shape the Saturday mornings of American children for generations to come.

Impersonations and Impersonators
I have been born more times than anybody except Krishna,
I suppose.
—Mark Twain

June 21, 1992. I close the door to the study. I put on a *Gipsy Kings* tape to block out the noise from the kitchen, where I know the family will be occupied for a while, Sunday being a favorite television night. I turn on my computer and settle down for some serious productivity. I am deep into my new book on Mark Twain and have made it clear that I am not to be disturbed.

My ten-year-old bursts into the room: "Mom, come watch *Star Trek* with us!"

"Can't you see I'm working?"

"But Mark Twain's in it!"

"Nice try. Some other time, sweetie."

A half hour later I wander into the kitchen for a cup of coffee. Mark Twain is chatting amiably with Whoopi Goldberg on the TV screen.

"That's—it's—why didn't somebody come get me?"

My son grins. I settle down next to him to watch the end of Part 1 of "Time's Arrow" on *Star Trek: The Next Generation.*

He is everywhere: on TV, in the flesh at shopping malls and casinos, in print as a detective or even as a love interest. Twain impersonators began to surface well before Twain's death. Much to his dismay, they tended to be petty crooks angling to get out of paying their own hotel bills. An occasional fraud with access to a printing press sought to cash in on Twain's reputation during his lifetime. He was infuriated by a "Will Clemens"—a remote relation at best—whose books played up an implied connection. The first Twain impersonation in print may have been a macabre 1906 short story in a Romanian journal written by one Vasile Pop and purporting to be an interview with the famous author. One wag, writing in 1909 (a year before Twain's death) argued in the *North American Review* that Mark Twain had really died three years earlier and that the author of *Christian Science* and *Is Shakespeare Dead?* was not, in fact, Mark Twain but the didactic writer Elbert Hubbard impersonating him. But what has become a steady stream of Twain impersonators in print really began seven years after his death with the publication of a book worthy of mention in a "News of the Weird" column.

Jap Herron, published in 1917, presented itself as a novel dictated by Samuel L. Clemens from the spirit world, via a device resembling a Ouija board, to a woman named Emily Grant Hutchings. On some level, of course, Mark Twain had it coming to him, since his own novels provided a writer's handbook of sorts for just such scams. After all, his novel about Joan of Arc was written not by Samuel Clemens *or* Mark Twain but by "the Sieur Louis de Conte," whose "personal recollections" of the Maid were supposedly "translated from the ancient French" by a "Jean François Alden." And *Adventures of Huckleberry Finn* is the work of none other than Huck Finn, of course. And then there were the diaries Twain found, the ones Adam and Eve kept. . . .

Hutchings plays it straight in her preface, "The Coming of 'Jap Herron,'" explaining that Samuel Clemens selected her as the conduit through which he would dictate, via "a planchette and a lettered board," two stories and a novel he had neglected to write while on earth. Straining to impart verisimilitude to her preposterous premise, Hutchings doggedly elaborates on the transcribing sessions in forty-two pages of plodding prose. As a publicity gambit the experiment was a failure. Even a putative assist from Mark Twain couldn't garner much attention for a novel this weak (a cursory search for reviews turned up none).

Despite his stance of public skepticism toward spiritualism, Twain seems to have kept an open mind. A member of the English Society for Psychical Research for a number of years, he read their publications with interest. He and his wife even visited a London medium on one occasion in an unsuccessful effort to make contact with their beloved daughter Susy after her death. This openness to psychic phenomena may have encouraged attempts to "make contact" with Twain after 1910. In any case, he seems to have made something of a habit of speaking from the grave. Posthumous "dictations" from Twain by planchette are "recorded" in Eunice Winkler's "Return of Mark Twain," published in 1918, and in *God Bless U, Daughter,* by "Mildred Burris Swanson and Mark Twain," which appeared in 1968. In one case Twain apparently chose not to speak through an unconventional medium but to opt for reincarnation instead. Michael Frank, an editor at the Mark Twain Project, recalls that in the 1960s a man repeatedly pestered the mystified staff with letters and phone calls about whether they had anything with Twain's fingerprints on it. Finally they dragged it out of him: he was convinced that *he* was Mark Twain and wanted to check his own fingerprints against Twain's for confirmation.

If Twain was prolific after death, H. M. and D. C. Partridge, authors of *The Most Remarkable Echo in the World* (1933), argued that he was infinitely more prolific in his lifetime than anyone had imagined. In this bizarre volume they painstakingly "proved"—through techniques including close reading of texts, minute examination of illustrations, cryptography, hermeneutics, etymology, numerology, symbology, graphology, and dactylology—that Mark Twain actually wrote works generally attributed to Edgar Allan Poe, Nathaniel Hawthorne, and Lewis Carroll. With the determination of those who play Beatles records backward to decipher the hidden messages they contain, the Partridges urge us to view Twain's texts and the accompanying illustrations *very carefully* in order to extract the coded secrets buried within. Some of the "secrets" are actually there to be seen with the naked eye. For example,

> The picture on the opposite page entitled "The Political Pot" is taken from Mark Twain's *Following the Equator,* Chapter LXVI. "Poe" is discernible in the word "monopoly" written in the flames issuing from the cauldron. On the ninth burning log from the left, is "Gold Bug," and on the seventh burning log, "negro." Both are placed directly

beneath "Poe." A negro slave is one of the principal characters in ["Poe's"] *The Gold Bug*.

The Partridges' observations, here as elsewhere, are accurate and precise; the conclusions they draw from them, however, are demented.

During the eighty-six years since Mark Twain's death, print impersonations and appropriations of his life and work have developed into a flourishing industry that attracts both amateurs and professionals. Some books present themselves as written by Mark Twain or one of his characters, while others feature those characters or Mark Twain himself in starring or supporting roles. There are books that self-consciously rework his plots, and there are books in which Twain manuscripts, artifacts, or houses in which he lived play a central role. A far from systematic survey turned up well over thirty texts that fit one or more of these categories.

Although Hutchings' *Jap Herron* may have been the strangest book "written by Mark Twain" after his death, it was by no means the most audacious. While one had to believe in Ouija boards for Hutchings' premise to make sense, it required no more than the ordinary reader's trust in the integrity of men of letters to be duped by *The Mysterious Stranger, A Romance*, which appeared in 1916. Presented as a previously unpublished book by Mark Twain, the volume was actually an editorial fraud perpetrated by Albert Bigelow Paine, Twain's official biographer and literary executor, in cahoots with Frederick A. Duneka of Harper and Brothers, Twain's publisher. Paine and Duneka took the unfinished manuscript of "The Chronicle of Young Satan," deleted sections that might offend Catholics or Presbyterians, cut and pasted until some coherence (by their lights) emerged, inserted the occasional character that Twain had developed elsewhere, borrowed the ending from another unfinished manuscript, and otherwise bowdlerized, presumably to preserve Twain's reputation. This "spurious version was delivered to an unsuspecting public in the form of a children's Christmas gift book," writes William M. Gibson, who edited the 1969 Mark Twain Papers volume, *Mark Twain's Mysterious Stranger Manuscripts*. Why? Gibson believes that Paine and Duneka's desire "to get out another book by 'Mark Twain' " played a major role.

The same desire was undoubtedly part of what motivated freelance writer and editor Charles Neider to "improve on" several of Twain's books himself and present the finished product as Twain's. In 1985 Neider, who has edited a dozen books on Mark Twain, published a version of *Huckleberry Finn* that condensed and cut vast chunks of the last third of the novel. The author, Neider believed, didn't *really* mean to have Tom Sawyer take over Huck's novel as he did, so Neider reined him in. Neider also removed what he felt were the boring passages from *A Tramp Abroad* in his 1977 edition of that volume and published Twain's chapters of *The Gilded Age* separately from Charles Dudley Warner's in a book he titled *The Adventures of Colonel Sellers* (1965). In each case Mark Twain is presented as the author of Neider's cut-and-paste job.

Mark Twain has taken time out from his posthumous writing projects to turn up in at least a dozen novels, sometimes as a central character, sometimes in a cameo appearance. They include historical novels, romances, science fiction, detective fiction, and young adult books. They are set in the past, the present, the future, or some combination of the three. Twain plays a rather dizzying range of roles in these novels, where he is reincarnated in youth, in middle age, and in old age. He does things we know Twain actually did, things he never did but might have done if he'd had the chance, and things that he could only do in the imagination of a writer born decades after his death. These books have little in common save the presence of Mark Twain and the author's efforts to get his "voice" right. Each author takes his or her liberties with Twain's character, some more plausible than others. Each challenges us to consider what we are willing to imagine: Twain leading an army? having an affair? courting a prostitute? Vertigo results if we try to meld the myriad personae that emerge from these books into one coherent figure.

The first time Sam Clemens/Mark Twain played a major role in a novel he didn't write was in 1971 in *The Fabulous Riverboat*, the second volume of Philip José Farmer's Riverworld series, a body of work that has been likened to Isaac Asimov's Foundation sequence and Frank Herbert's Dune books in importance, ambition, and scope. Set on a ten-million-mile-long river, the novel merges aspects of *Extracts from Captain Stormfield's Visit to Heaven*, the last book Twain published during his lifetime, with dimensions of *Huckleberry Finn, Life on the Mississippi, What Is Man?, Tom Sawyer Abroad*, the "Mysterious Stranger" stories and, most of all, *A Connecticut Yankee*. Farmer shows himself to be a careful as well as imaginative reader of Twain's work, evoking not just the man's voice but also themes that fascinated him: the mixed blessing of "progress", the nature of dreams; the psychological impact of slavery; the seductions of technology, the danger and challenge of the river. Farmer's Clemens is required to respond to turns of fate that the real Sam Clemens never had to deal with, some more fanciful than others. How does he react to the charge that some consider *Huckleberry Finn* a racist book? With a groan, after which he elicits the confession that the person making the charge has never read the book. How does he respond to the fact that his supposedly scandalous posthumous publications caused relatively little stir? With shock. To the fact that his daughter Clara has converted to Christian Science? With amazement. To the news that in her next life his beloved Livy has taken up with Cyrano de Bergerac? With a tirade against "that dirty, uncouth, big-nosed, . . . vigorous, talented, scary Frenchman."

Sam Clemens was resurrected several years later by Gore Vidal, who gave him a small role in his 1976 novel, *1876*. As usual, Vidal did his homework; Twain's conversation over drinks at Delmonico's is plausible both in content and in style. For instance, speaking of the theatrical flop by Bret Harte that he has just witnessed, Twain says, "And to think that my old friend wrote that thing, or, I reckon, just let it happen to him, like measles." Twain is still drinking in Delmonico's in 1982 in Joyce Carol Oates's novel, *A Bloodsmoor Romance*. Despite the real Mark Twain's legendary fidelity, the fifty-eight-year-old fictional Twain we meet in this feminist historical romance is captivated by a thirty-something actress named Malvinia Morloch, whom he takes as his mistress at a time when he is plagued by financial troubles and by the ill health of his wife (who has remained in Europe) and of his daughter Jean. Oates's Twain sounds nothing like Farmer's or Vidal's:

> "What is Man?" the aging, but still handsome, gentleman rhetorically inquired, sprawled in a booth at Delmonico's or at his specially reserved table at Sherry's. "What are *we*, my dear Miss Morloch, but machines? No, no, pray do not turn so startl'd and innocent a visage upon me, my dear, but only consider: machines moved, directed, and commanded by *exterior* influences, *solely*—or, if animal-like, exactly along the ignoble lines of the chameleon, who takes his color, in order that he may survive, from his place of resort."

But Oates, like Vidal, has researched her material. Mark Twain did spend a good deal of time in New York in 1894, during his fifty-eighth year (while Livy remained in Europe), and although *What Is Man?* would not be published for a dozen years, he claimed to have been working through the ideas in it since the 1880s.

In 1985, a hundred and fifty years after his birth, seventy-five years after his death, and a hundred years after the publication of his greatest novel, Halley's Comet miraculously transports a living, breathing Mark Twain back to earth in David Carkeet's *I Been There Before*. The novel has become a cult classic among Twainiacs. Puzzled to find himself tramping barefoot through a snowy landscape in an indeterminate location, Carkeet's Twain speaks in tones that are deliciously familiar.

> "This is good," I thought. "A road is good. A road will lead somewhere. That is what roads are for." My thoughts were not sublime. They came singly, each one simplicity itself, and each stuck in my brain for longer than it deserved. Perhaps this is the way idiots think. My next thought was this: What did I last remember? My bed—in Stormfield. Pain in the breast—that curious, sickening pain. . . . The recollection came as a comfort, for of course I was dreaming. I was on my deathbed in Stormfield dreaming I was tramping around this barren, snowy landscape, dressed only in my night-gown. I neglected to mention this last fact earlier, out of modesty, but there it is now: a night-gown.

Particularly memorable is Clemens' incognito visit to the Mark Twain Papers, "quite a busy little beehive across the Bay, & *I'm* the honey that all the buzzing is about," as he describes it in a letter to Livy. The young woman who invites him in asks whether he is a Mark Twain scholar.

> "Hmmm," I thought . . . "No," I said.
> "A buff?"
> "Hmmm" again. My thoughts drifted to buffalo skin, to light yellow. I warned myself not to be a ninny. I hung fire.
> "A fan?"
> I wasn't at all warm, so *this* offer puzzled me.
> "Hello?" she said, to see if my spirit was still present, I suppose.
> "Hi!" says I, hoping this would put us on a modern & friendly footing.
> She frowned. It didn't suit her soft face. She asked as a last resort, "Do you have any favorites among his works?"
> "*Joan of Arc*?" Livy, I don't know why I *asked* it like that, but at least I got that frown off her face—now she was grinning. I was mighty pleased with myself.
> "That's refreshing," says she.

Whereupon she takes him to the "narrow room with a long empty table facing the windows"—familiar to all who have worked in the Mark Twain Papers—to wait while she fetches the manuscript. "I think Clemens tried to be something he wasn't when he wrote *Joan*," she muses, leaving the box containing the manuscript on the table.

"He was tired of being the buffoon, so he tried to be high-toned & serious. The result was a very boring book. . . ."

"He loved this work," I said, feebly.

"Even Howells panned it."

I sighed. "I know," I said.

In 1987, two years after *I Been There Before*, Kirk Mitchell's *Never the Twain* appeared. Howard Hart, an impecunious descendant of Bret Harte, is visited by a scholar who lays a curious theory on him: if Clemens hadn't bungled the ownership of a silver mine that eventually produced millions, financial necessity wouldn't have forced him to become a writer, Bret Harte would have remained the reigning presence in Western literature, and Howard Hart would have inherited what would have become Harte's multimillion-dollar estate. Howard happens to have a wild-eyed friend named Roderigo who has a bold theory about time travel and has been lobbying him to try it out. Howard finally agrees and goes back in time to the Comstock Lode in the spring of 1861 to make Twain rich. He doesn't succeed. But along the way, while skillfully interlarding passages from *Roughing It*, Mitchell captures the Nevada mining country in well-researched detail, melding historical accuracy with futuristic fantasy to spin a witty tale of greed and romance, frustration and thwarted revenge.

Similarly innovative and entertaining is Allen Appel's *Twice Upon a Time* (1988), a fast-paced adventure story that blends history and science fiction and also features travel to the past. During the Philadelphia Exposition of 1876, Twain's fascination with a bit of modern technology that the hero, historian Alex Balfour, has unwittingly brought with him provides precisely the opportunity for intervening in the past that Howard Hart had longed for, but Alex finds the prospect terrifying.

"And what the hell is that?" [Twain] said, bending over and peering at Alex's zipper. "I noticed that damn thing when you came in this morning. The man's coat has no buttons, I said to myself. Just a curious strip of perforated metal up either side of the front. Brass, isn't it?"

"Uh, yes. I guess," Alex said. He suddenly had a vision of the changed future: Mark Twain, Inventor of the Zipper. No *Huckleberry Finn*. No wellspring from which Hemingway would claim the roots and greatness of the modern American novel. . . . Twain tugged the zipper a few inches down, then back up. A beatific look came over his face. . . .

William Dean Howells, P. T. Barnum, General George Armstrong Custer, two slimy con men named King and Duke, and a black storyteller who bests Twain at his own game all figure in a plot that spins increasingly out of control but that can hardly be accused of straining credulity any more than Twain's *Connecticut Yankee* does.

In Darryl Brock's 1990 novel *If I Never Get Back*, a modern-day newspaper reporter from San Francisco named Samuel Clemens Fowler accidentally lands in 1869 and meets his namesake, who ropes him into a get-rich-quick scheme that involves grave robbing in Elmira and a violent tangle with the Fenians. Clemens indeed had a weakness for moneymaking schemes, but not if they were dangerous or of questionable legality. And there was also no time in his life when Clemens would have been less disposed toward such a scheme than on the eve of marrying into the wealthy Langdon family, when he was trying to reassure his future in-laws about his respectability and trustworthiness. More believable is Clemens' motivation for setting out after hidden treasure in Peter J. Heck's *Death on the Mississippi* (1995). Heck has a financially troubled Twain sign on as a lecturer on the *Horace Greeley* and travel down the Mississippi in the 1890s. The ship's passengers include assorted crooks, thugs, and cardsharps, but the one person from whom Twain recoils in genuine terror is the aspiring author of an epic poem about the Civil War.

Late-twentieth-century journalism and fiction has kept Mark Twain's characters almost as busy as their author. Huckleberry Finn was particularly active, turning up in 1986, for example, with an assist from Roger Rosenblatt, as author of a column about freedom in *Time* magazine on the occasion of the hundredth birthday of the Statue of Liberty. And he surfaced again in 1993, reviewing *Was Huck Black?* in the *American Historical Review*, with some help from Karen Lystra.

> That book was made by Mrs. Shelley Fisher Fishkin, and she told the truth, mainly. . . .
>
> Mrs. Fishkin got out a heap of books and learned me about Jimmy, a black child who nearly wore Mr. Twain's ears off in some fancy hotel one time. Had a powerful influence. I was in a sweat to find out all about him. She says we talk alike. Repeating words, and tangling them all up, and unfixing their meanings, and using something she named a parcipial construction and ten shifters and castrated conjunctions and non-redundant tense markings and cereal verbs and double on the negatives and just sort of piling it on to beat the band.

In between stints as a contributor to magazines and scholarly journals, Huck also spent a fair amount of time fiddling with his first book—

although, as he said of the food at the Widow Douglas's table, "there warn't really anything the matter" with it.

In the introduction to *The true adventures of Huckleberry Finn* as told by John Seelye (1970), Huck tells us how one twentieth-century "crickit" "got right to work down in the innards of [*Huckleberry Finn*], and showed how sloppy it was put together. He would tear a part out and show how loose it had been wired in, and then he would reach down and tear out another part. It was bloody but grand." All the critics' complaints galvanize Huck to take action: "So I thought to myself, if the book which Mr Mark Twain wrote warn't up to what these men wanted from a book, why not pick up the parts—the good ones—and put together one they would like?" Huck writes, "And now that they've got *their* book, maybe they'll leave the other one alone." In the postscript Huck attaches to the second edition of the book, published in 1987, he includes a report on a Twain symposium held in Missouri in 1985. One of the "crickits" at that "confrince" was Dr. John H. Wallace, whose version of *Huckleberry Finn* changes all the occurrences of the word "nigger" to "slave" throughout and also omits anything that thereby ceases to make sense, like Huck's famous exchange with Aunt Sally. The substitution fails to impress Jim, Huck tells us: "Jim said that was plain silly. 'In de fus' place,' he said, 'how is I gwine ter be a slave when I 'uz already run off? En' in de second, how is I gwine ter be a slave when ole missus already done set me free?'"

As is well known, Mark Twain developed a fondness for his characters and had no objections to milking a good thing; he brought Huck and Jim and Tom back for subsequent adventures—with decidedly mixed results. Sequels have also been attempted by other writers, but most often they drive us back to the originals to reassure ourselves that our old friends are still there for us and haven't somehow been flattened into these thin creations of lesser talents. Reading Clement Wood's *Tom Sawyer Grows Up* (1939) or *More Adventures of Huckleberry Finn* (1940), for instance, is a bit like reading a comic-book reduction of a Cliff's Notes version of Mark Twain.

Greg Matthews, an Australian writer now living in the United States, fares better in *The Further Adventures of Huckleberry Finn* (1983), boldly writing in Huck's voice for no less than five hundred pages—a task that would have given Twain's Huck the fantods. While Matthews' Huck may not have as much grit and wit as Seelye's, he's recognizable: he often sounds like the Huck we know. We learn that freedom agrees with Jim, who has bought his wife and children out of slavery and lives with them (rather improbably) in St. Petersburg at the Widow Douglas's house, where he's employed. But while Jim is off with his first jug of whiskey celebrating the miracle of his daughter's unexpected recovery of her hearing, the widow's house burns down and he is left with no family at all—with the exception of Huck, that is, who breaks the dreadful news to him the next day. For this reader, at least, not even 441 new pages of Huck and Jim hunting for gold and meeting up with Indians, gamblers, whores, and gospel singers justify inflicting this new tragedy on an old friend who's already been through *enough*.

In 1986 *The Boys in Autumn*, by Bernard Sabath, opened on Broadway. Tom and Huck, "America's foremost symbols of boyhood optimism, of romantic dreaming at the age when all life's options lie open," wrote William A. Henry III in *Time*, are here transformed into "disillusioned middle-aged men . . . sputtering mistrust." After seeing the show, Henry noted,

> many appalled theatergoers demanded, Who is Bernard Sabath and how dare he defile these golden scamps? That outrage underscores how deep a nerve the playwright is aiming for. If life has so disappointed Huck and Tom, the epitomes of hope, how can a spectator not be plunged into pessimism about his own unfulfilled ambitions?

Unfortunately the play's "execution [was] not quite so imaginative as the premise."

Mark Twain's books and manuscripts have a mysterious habit of turning up in twentieth-century murder mysteries. Much of Lee Thayer's *Murder Stalks the Circle* (1947), in which a faked piece of Twain memorabilia plays a key role, is set in the old Mark Twain house on Farmington Avenue in Hartford, where Thayer herself spent the summer of 1945 when it was a boardinghouse and public library. Among the several murders committed in the book, one that would have been sure to gall Mark Twain was that of a cat—struck down in his own house. ("A home without a cat—and a well-fed, well-petted and properly revered cat—may be a perfect home, perhaps, but how can it prove title?") A previously unknown but presumably authentic Twain manuscript entitled "The Death of Huckleberry Finn" is central to Yvonne Montgomery's absorbing *Scavengers* (1987). Another 1987 mystery, Julie Smith's *Huckleberry Fiend*, features what has all the marks of being the long-lost first half of the manuscript of *Huckleberry Finn*, which in fact turned up in California four years later, in 1991. Murder and mayhem ensue, and before the fiend or fiends are uncovered, Smith's characters (sometimes using aliases with a familiar ring like "Sarah Mary Williams") engage in a fair amount of lively and convincing "Twain talk." Stolen letters supposedly written by Mark Twain are central to Sharon Heisel's young adult mystery *Wrapped in a Riddle* (1993), in which eleven-year-old Miranda plays sleuth during a visit to the Jumping Frog Inn, a bed-and-breakfast establishment run by her grandmother. Mark Twain is the subject of a prize-winning essay by a student murdered on page four of Edith Skom's *Mark Twain Murders* (1989), but despite its title the book has almost nothing to do with Twain and the "Twain talk" never rises above the pedestrian.

Mark Twain's manuscripts and books also surface periodically in genres other than mysteries, sometimes serving as a springboard for serious explorations of such issues as freedom and censorship. Howard Fast's *Silas Timberman* (1954) concerns the difficulties the hero experiences in the McCarthy era—all of which begin when he makes Twain's story "The Man That Corrupted Hadleyburg" the focal point of his American literature class. The plot of *The Day They Came to Arrest the Book* (1982), Nat Hentoff's brilliant young adult novel about censorship, is set in motion when a high school librarian refuses to follow her principal's orders to take *Huckleberry Finn* out of circulation. On the frothy side, Robert Guntrum's *Great Twain Robbery: A Comedy Caper* (1995) centers on an antiques-gallery heist that targets a priceless Twain manuscript.

Twain's plots have inspired an equally varied array of imaginative efforts. In Gwen Davis's *Princess and the Pauper: An Erotic Fairy Tale* (1989), a bored princess from the kingdom of Perq in the Irish Sea trades places with a Welsh housewife who looks remarkably like her; the two full-grown, libidinous, and married heroines add a new dimension to a plot borrowed from Twain. Science fiction grand master Isaac Asimov credits *A Connecticut Yankee* with being the first book "in which the hero or heroine has gone back into the past," a plot that has become a staple for "modern writers who have attempted science fiction or fantasy," many of whom are indebted to Twain's novel.

A Connecticut Yankee was cheerfully plundered again in 1989, the centennial of its publication, in a contest sponsored by the *Hartford Courant*: submit a five thousand-word "modern slice" of the novel and win publication in *Northeast*, the *Courant*'s Sunday magazine, and a first-prize trip for two to Camelot, or at least to England, "where Camelot may have been." More than two hundred writers from "most towns in Connecticut and from 12 other states" accepted the challenge. Their submissions were reviewed by a distinguished panel of judges that included Pulitzer Prize–winning Twain biographer Justin Kaplan. Mike Wavada, a forty-one-year-old software consultant who won first prize with "Sir Consultant's Strategic Plan" had a Hartford advertising copywriter land in Camelot when his BMW crashes into an oak tree in a dense fog. Greeted by men on horseback in rusty chain mail, he decides it must be "a convention of the Society for Creative Anachronism." But the men fail to laugh at his jokes, stubbornly remain in character, and seem distressingly hostile. He stuns them into submission by turning on his one working headlight. Before long he's solved Arthur's Big Problem (too many nobles collecting taxes and not enough people to pay them) by coming up with a clever plan that culminates in Joust-a-Mania, a major sporting event at which the concessions alone do much to fill Arthur's sagging coffers (a nonalcoholic cherry drink called "Sblood" is quite popular with the children). In "A Return to Camelot" by an eleven-year-old entrant, a Hartford computer repairman finds himself transported to Arthurian England, where he meets up with a Connecticut yankee who arrived there a hundred years earlier and has been running the place ever since:

> When I came to, he was staring at the contents of my attaché case, which he had strewn all over the floor. After asking me a few questions, like "What is it?" and "What's it for?" he concluded that by 1989 electronics was at least three times more complicated than in his day. . . .
>
> He said . . . "I'm posing as a wizard to make sure people are respectful. My 'magical powers' consist of gunpowder, electric lighting, dynamite, and plumbing. . . . What are your, um, magical powers?" . . .
>
> "In this bag is my laptop." Without waiting for him to ask "and what does it do," I told him. . . . Hank seemed lost. . . . I tried to state what I meant in simpler terms. "It is run by electricity, the screen looks like a square flat lamp, and you press keys on what's called a keyboard."
>
> For once something got through to him. In an elated voice, he said, "A keyboard! With one key for each letter! My, the typesetting machine I bought stock in must be doing well. I must be a rich man by now!" I hated to tell him. . . .

The sixth-grader's story "won the hearts of the *Northeast* staff," who published a long excerpt of it in the magazine (The name of the "Youngest Yankee," by the way, was Joey Fishkin. And I was one proud mama).

For every book featuring Mark Twain or one of his characters or one of his manuscripts or one of his plots, there are probably at least a dozen efforts to portray the man "in person." "The mortal Twain" would doubtless have been amused by the two impersonators "who resurrected him for Charlotte, North Carolina, on the same day in 1980," writes Louis Budd. Many Twain impersonators and actors who portray Twain (they often have strong views on which term is correct to describe them) are men who have other jobs in "real life." Henry Sweets recalls the firefighter from St. Louis who turned up in Hannibal to take some publicity shots as "Mark Twain." The photo session was long and the day was hot; the man's makeup began to melt like an ice-cream cone as he haplessly became himself again in the midday sun. In Memphis an assistant director of Parks and Recreation and a member of the county sheriff's staff both took a turn at "Twaining" at civic functions in 1994. Occasionally part-time Twain impersonators get bitten by the bug and decide to ditch "real life" altogether. That's what happened to Marvin Cole in 1994. Then president of DeKalb College, the third-largest college in the Georgia university system, Cole resigned his job and dedicated his life to impersonating and studying Mark Twain. He was proud that during his thirteen years as president DeKalb doubled its enrollment and employed more minority faculty than any other college in Georgia, but he was also committed to the higher education Twain was capable of providing. "I believe so much in what Mark Twain was trying to do and say," Cole told a reporter. "He said we are hypocritical people who don't know how to live with each other. I don't think we've progressed over the years in learning to become humane, and because of the increased diversity in our country, if we don't learn how to live together, we can become another Bosnia. I guess my act serves as both entertainment and a warning."

No two Mark Twain impersonators are alike, a fact that speaks volumes about the multifaceted nature of the man who wrote their acts. Some performances focus on Twain's later years, some on his youth, while others pull in riffs from all over his opus. Whichever Twain they choose to project, these actors share a deep love of the incomparable material that is theirs to work with. Twain impersonator F. X. Brown of Elmira, New York, observed that he had "one of the best writers in the world" writing his act and all he had to do was "learn the timing."

Ken Richters never planned to "become" Mark Twain. The Connecticut actor, who tours in his own one-man show billing Twain as "America's First Stand-Up Comedian," remembers driving cross-country from Hollywood to the East Coast and deciding to take a detour when he saw a sign for Hannibal on the highway. As soon as he got there, his car broke down. (He is hesitant to call it fate: the car also broke down in Phoenix.) During the three days he spent waiting for a new camshaft, he had a chance to immerse himself in the town, its history, and the folklore surrounding its most famous native son. He thought it might be fun to assemble a stage show of Twain material to perform when he made it to New York. His first shows were designed for high school audiences and focused on the irreverent Twain, the antiauthoritarian, antiestablishment Twain that Richters himself would have found appealing when he was in high school. He would make the first thirty-five minutes of the forty-minute show as funny as he could; then, in the last five minutes, he would "get deep with them, somewhat philosophical," and they would pay attention "if for nothing else in respect for the last half hour of laughter that I gave them." Richters performed Twain in over 150 high schools in 1980 and 1981. After that he began playing colleges, theaters, and corporate gatherings. Going on nine hundred performances later, he still emphasizes the sarcastic Twain, the social and political commentator, the man who intentionally courted controversies and got them.

McAvoy Layne, of Incline Village, Nevada, who has performed as Mark Twain in more than a thousand venues from his home state to the former Soviet Union, decided to make those controversies an integral part of the educational experience by letting students put Twain on trial for being a racist—a charge of which he tends to be acquitted after mounting his own defense. Layne finds that these "trials" provoke heated moral debates about racism, protest, satire, and censorship. He has impersonated Twain full-time for five years in his own shows, in television productions, as a greeter in the Silver Legacy Casino, and at trade shows as the alter ego of "Virtual Mark Twain." What sustains his interest in Twain? Layne answers by quoting Justin Kaplan: "In the end the man is more imposing than the sum of his works." Layne treasures a flyer autographed by Hal Holbrook: "McAvoy—keep the torch burning."

Holbrook is without question the doyen of Twain interpreters. Portraying Mark Twain onstage in countless performances has not dimmed his enthusiasm for the man or his work. Holbrook finds that as he himself has aged, his understanding of Twain has changed as well, always yielding an inner wisdom that repays his attention. In 1954 Holbrook immersed himself in Twain's work to create a one-man show. His painstaking research set a Himalayan standard against which other Twain interpreters would be judged (and usually found wanting). From the start, Holbrook recognized the challenge posed by the absence of any recording of Mark Twain's voice. He calls his act a portrayal rather than an impersonation in part because of the guesswork it entails. "What did he sound like?" Holbrook asked himself more than forty years ago.

I had read newspaper reports of his lectures which described his voice as a "nasal twang" or "a little buzz saw inside a corpse," and the *Detroit Free Press* had referred to his "Down East" accent. Others called it a Missouri drawl. I learned early on in my research for the solo performance of Mark Twain that I would have to deal with inconsistency as well as the creative flights of the people doing the reporting. Some of them were trying to be funnier than Twain. And there was always the possibility that the man on the *Detroit Free Press* did not know a Down East accent from a Paiute Indian's.

Holbrook studied accounts of Twain's lectures and sought out people who had known him; he was even granted an interview by Twain's secretary, the elusive Isabel Lyon. He gave his earliest Twain performances at Upstairs at the Duplex, a Greenwich Village nightclub, with a cigar as his only prop.

I worked in the curve of a baby grand piano, very close to the audience, and did two or three shows a night. By then the makeup took two and a half hours. I stayed there for eight months and developed my first two hours of Twain material. I experimented with the pause. Twain loved fooling around with the pause, toying with it like a cat, taking chances with it, and there in the nightclub I was able to get familiar with the pause, especially in the late show when no one was in a hurry to go anywhere, the smoke drifting up through the spotlights and people simmered down from a nice load of whiskey or gin. I could take my time telling the stories, let the stuff sink in. . . . I learned to trust Twain's material.

Holbrook's enormous respect for Twain's words has endeared him not only to Twain fans around the world but also to Twain scholars, who honored him in 1996 by making him an honorary member of the Mark Twain Circle of America. While lesser talents have no qualms about putting unlikely words in Twain's mouth, the more than twelve hours of monologues Holbrook has committed to memory are almost pure Twain. The results, more often than not, are electric. Holbrook's extraordinary command of his material allows him to tailor each performance to both the audience and current events. Despite the appearance of spontaneity, which Twain himself carefully cultivated, nothing is improvised; everything is planned.

Holbrook did not become fully aware of the potential for social commentary in Mark Twain's work until 1957. Prior to that time he had simply been striving for laughs.

But when President Eisenhower called out the troops to put down the racial explosion at Central High School in Little Rock, Mark Twain's social conscience began to cast its shadow over me. By some strange twist of fate (later to be repeated in the 1960s) I was scheduled to perform Mark Twain near Little Rock not long after the riots there. I did not yet have material in my repertoire that specifically commented on racial injustice. All I had was the "Sherburn-Boggs" selection from *Huckleberry Finn* which ends with Colonel Sherburn's blistering speech to the mob that has come to lynch him. Although a white man is speaking to a white mob Mark Twain is making a thinly veiled statement about the Ku Klux Klan. The portrait of sudden violence in the shooting of Boggs, of ignorance and the mob mentality that sweeps people along, was eerily appropriate to this modern-day crisis, and Twain's setting did happen to be a town in Arkansas. So that was the selection I chose to deliver.

From that point on, Holbrook paid close attention to Twain's social commentary. "Suddenly the show took on a feeling of importance it had never had before. I didn't know it then but the Civil Rights Movement had begun." The highest compliment he ever got, Holbrook says, "was doing Mark Twain in Oxford, Mississippi, after the riots, when James Meredith was there."

"Mark Twain Tonight" was the first thing whites and blacks were allowed to assemble for; I was the guinea pig. I expected the show to be canceled but it wasn't, so I went out and did the best stuff I had on civil rights and human justice. Afterwards a lady of about 80 or 90 years came backstage and she said, "Mr. Holbrook, you just gave a better sermon tonight than we heard in church last Sunday."

During the heat of the Vietnam War, Holbrook moved from racism to patriotism, using Twain's comments on America's behavior in the Philippines. The material proved particularly powerful, reminding the audience that "dissent is a tradition of our democracy, something we forget about six and a half days of every week."

Holbrook's meticulous, thoughtful, imaginative, deeply engaged and deeply engaging interpretations of Twain's words over a forty-year period before hundreds of thousands of people have given Twain the one thing he could not give himself: a vitality beyond the grave that no author has the right to expect.

16. *Actor Hal Holbrook in his legendary one-man show, "Mark Twain Tonight"*
(Photo courtesy of Hal Holbrook)

Technology

The trouble with these beautiful, novel things is that they interfere so with one's arrangements. Every time I see or hear a new wonder . . . I have to postpone my death right off.
—*Mark Twain*

November 1995. I had fed the boys breakfast, sent them off to school, and changed out of my bathrobe into clothes in a pre-caffeine haze. I sat down to my first cup of coffee. The doorbell rang. I glanced out the window at the pleasant-looking young man with a briefcase who stood on my doorstep. It was too early in the day for Jehovah's Witnesses or Greenpeace canvassers, and it was past the time of year when local political aspirants made their door-to-door rounds. Then it hit me: the man from IBM. The one who phoned last week. The one the New York office had sent to call on me. "You must be Brad Messerle," I said, opening the door. "Thank you *so* much for coming!" As I watched him set up his laptop on my kitchen table, my eyes met those of Mark Twain, who peered at me from a poster on the wall. You sure get me into some strange situations, I thought, as I returned Twain's gaze: Here I was, someone who hadn't yet penetrated the mysteries of programming her VCR, about to have a personal demonstration of technology so new even my "techie" friends hadn't heard of it yet. . . .

Brad Messerle was not the sort of man who played music to keep his plants happy. But there he had sat, in the fall of 1994, in his San Antonio living room, reading a short story by Mark Twain to his computer. "I took a large room, far up Broadway, in a huge old building whose upper stories had been wholly unoccupied for years until I came," he read. "The place had long been given to dust and cobwebs, to solitude and silence. I seemed groping among the tombs and invading the privacy of the dead, that first night I climbed to my quarters. For the first time in my life a superstitious dread came over me. . . ." He continued to read for an hour and a half, as he knew he must. Nothing else would work. It had to be that story by Twain. By the time he was done, his computer had learned to recognize his voice. The next day, when he arrived at his desk, he said "Start Dictation," and the computer transcribed onto a screen whatever came out of his mouth. With the touch of another few buttons his words were transferred cleanly and crisply to paper.

This scenario is not the fantasy of an out-of-control Twainiac. It is a simple statement of what happened the day Brad Messerle got his copy of IBM's new VoiceType Dictation system to work for the first time. This state-of-the-art dictation system, the product of over twenty-five years of research at IBM, requires that the user read two texts out loud: a tutorial overview of the product and Mark Twain's "Ghost Story." Utilizing a series of patented algorithmic equations to isolate, identify, and interpret the phonemes in human speech, the computer then builds a mathematical model unique to the user's voice which is used to interpret all acoustical data. Spoken words are displayed on a monitor and delivered to a word processor: the results are indistinguishable from text that has been typed.

When I first heard about the VoiceType Dictation system and the use it made of Twain, I was intrigued. Why Twain? After the technology was designed, IBM assembled a team of experts, including a psychologist, to choose a script for the training session. Since the selection had to be read slowly, a dull piece would make the entire experience seem interminable. The goal was to make the user interested enough to read the script through to the end. They finally converged on Twain's "Ghost Story." First published in the *Buffalo Express* in 1870 and later reprinted in Twain's *Sketches, New and Old* (1875), the piece is more a spoof of a ghost story than the real thing, but Twain throws in a fair amount of genuine suspense and eeriness as well, making the story part thriller, part send-up of thrillers, part humorous sketch, and part social satire. Once the user has finished reading "A Ghost Story" out loud, the computer is able to produce text ready for editing within seconds of hearing dictation. IBM claims that the system's advanced phonetic and linguistic tools allow it to understand dialect and strong accents and to distinguish among homophones in context (e.g.,"to," "two" and "too"). It can recognize the start of a sentence and provide capitalization, and is said to have a 97 percent accuracy rate.

Mark Twain would have loved it. He probably would have sunk a fortune into it if he had happened to have one handy. It was precisely the sort of technology he was always on the lookout for. It would have helped him indulge the preference for dictation over writing he developed in his final years, and would have appealed to his delight in techno-logical innovation. It would have captured his imagination and his bank-book, had he not already lost the latter by backing the doomed Paige typesetter.

In a workshop he rented from the Colt Firearms Factory in Hartford, Connecticut, a New York–born machinist named James William Paige de-veloped the prototype for an extraordinarily ambitious mechanical compos-itor and filed for a preliminary patent for it in 1872. He received a patent for "an Improvement in Type-setting Machines" in 1874, ten years before Mer-genthaler finished his first Linotype machine. The patent he requested for the finished machine some thirteen years later contained over two hundred sheets of drawings with over a thousand views of the machine's eighteen thousand separate parts. It took the patent office eight years to process, dur-ing which time the patent attorney who originally prepared the case died in an insane asylum, as did one of the patent examiners.

When it was working properly, the Paige typesetter was awesome: it could set twelve thousand ems of type an hour in beautifully justified lines. (Mergenthaler's Linotype machine could set just eight thousand ems an hour.) Twain was beside himself with excitement after the machine's first full-blown test run in 1889. He wrote his brother Orion,

> At 12:20 this afternoon a line of movable types was spaced and justified by machinery, for the first time in the history of the world! And I was there to see. It was done *automatically*—instantly—perfectly. This is indeed the first line of movable types that ever *was* perfectly spaced and perfectly justified on this earth. All the other wonderful inventions of the human brain sink pretty nearly into commonplace contrasted with this awful mechanical miracle. Telephones, telegraphs, locomotives, cotton gins, sewing machines, Babbage calculators, Jacquard looms, perfecting presses, Arkwright's frames—all mere toys, simplicities! The Paige Compositor marches alone and far in the lead of human inventions.

It was impossible to overestimate, Twain felt, the importance of printing technology to progress. Twain was right that "one of the hundred and one devices and inventions for setting type by machinery" would be the wave of the future. Unfortunately, that one device wasn't his.

In many ways the Paige typesetter and IBM's VoiceType Dictation system were analogous technological breakthroughs. While the Voice-Type Dictation system saved the user hours of keyboarding, the Paige typesetter saved the user hours of tedious hand typesetting. Despite their patented ingenious designs and great complexity, both took relatively little time for the user to master, and both required trained technicians to troubleshoot if problems arose (experienced machinists stood by ready to jump in and repair glitches whenever the Paige typesetter was operating, and IBM experts man a 1-800 hotline to address difficulties that may arise involving the VoiceType Dictation system.) And both produced the same end product: words on a page. There was one key difference, however. The VoiceType system (if I could believe the handful of IBM folks and users I'd interviewed) actually worked, while Paige's "wonderful type-setting machine" broke down so frequently as to ultimately be deemed useless.

Brad had booted up his PC by now. "Watch this," he said, taking out of his attaché case the kind of debriefing report a doctor would make after seeing a patient. Brad said "Start Dictation" to his laptop in a firm voice, and started reading from the sheet, enunciating clearly and distinctly. The words appeared on the screen as he read them, and even the

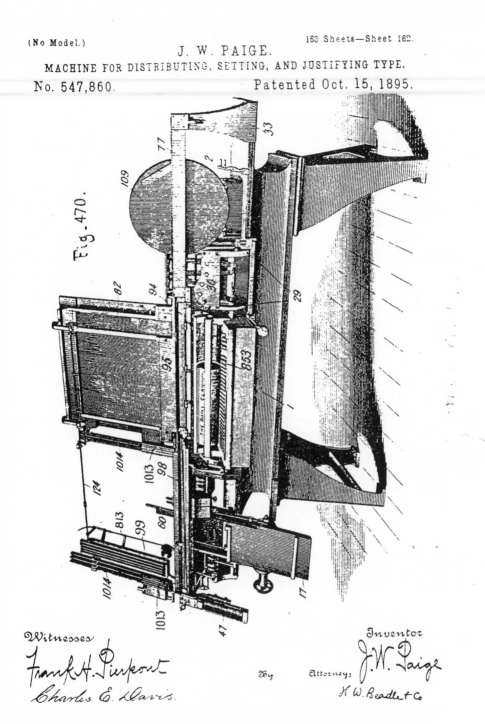

17. *Sketch of the Paige Compositor, the automatic typesetting machne that bankrupted Mark Twain (Photo courtesy of the Mark Twain House, Hartford, Connecticut)*

most complicated medication was spelled right. "This is one of the kinds of things it's best at," he said, noting that a large number of VoiceType Dictation systems had been sold to medical professionals. I asked if I could try it. "Sure," he said, "but don't expect it to work." I read from the same sheet of paper he had just been reading out loud, but, sure enough, the words failed to appear on the screen. Like a dog trained to respond to only its master's voice, the system recognized his voice and nobody else's (that was what he had trained it to do by reading it the Twain story).

Okay, so it would work only for him. But how well would the VoiceType Dictation system work if he read it something else by Twain, something trickier than "A Ghost Story," something like the first and last paragraphs of *Huckleberry Finn*? "Only one way to find out," he said. I fetched a copy of the book from the study. He said "Start Dictation," then proceeded to read the first and last paragraphs of the novel out loud in slow, clear tones. The "truth, mainly" came out as the "truth namely." "Anybody but lied" came out "any body butterfly." The "widow" came out as "window." "Stretchers" came out as "structures." Instead of being "rotten glad," Tom was "writing glad." "If I'd a knowed what a trouble it was to make a book I wouldn't a tackled it and ain't agoing to no more" got garbled into "If I'd a node what a total it was to make a book I want a tackled it had a according to no more." "Light out" became "live out" and "adopt" became "about." Huck's claim that Aunt Sally was going to "sivilize me" proved a real stumper for the program; it valiantly came up with the notion that she was going to "stabilize me."

So here was something else the VoiceType Dictation System and the Paige typesetter had in common: both were spectacular during the demonstration, but more complicated in practice.

Twain was intrigued the moment he first heard about Paige's machine from Hartford jeweler Dwight Buell, who cornered him in a billiard room; when he saw it in action, he was hooked. Paige, he proclaimed, was "a most great and genuine poet, whose sublime creations are written in steel. He is the Shakespeare of mechanical invention." After an initial small investment in 1880, Twain became Paige's partner in 1886, and began putting three thousand dollars a month into the project. His investment would eventually amount to some three hundred thousand dollars. When it worked, Paige's typesetter was more precise and impressive than Mergenthaler's Linotype machine, but Mergenthaler's machine proved more dependable. Paige's invention, which he continued to tinker with and never actually "finished," drove Twain into bankruptcy. Twain had had the prescience to enter the right race: automated type-setting *would* be the wave of the future; he had simply backed the wrong horse.

Twain had put about twenty-five or thirty thousand dollars into each of several previous business ventures, including a steam pulley, an en-graving process, a patented steam generator, and a new method of marine telegraphy. And he had long dabbled as an inventor himself, intrigued by the potential that inventions held as eventual financial bonanzas. In *A Connecticut Yankee* he would call Johannes Gutenberg, James Watt, Sam-uel F. B. Morse, and Alexander Graham Bell "the creators of this world—after God," and would later add Thomas Edison to the pantheon as well. He had a telephone installed in a booth off his main entry hall in Hartford in 1878, with a direct link to the *Hartford Courant* and later to the telegraph office, and boasted that it was the first private telephone in the world. He had a central gas furnace installed as well, claiming it was the first of its kind in a private home. He tried out such newfangled creations as the telharmonium, a device which transmitted music by wire from a microphone to an amplifier.

Twain himself invented a gadget to attach sheets and blankets to beds, a self-adjusting vest strap, a spiral hatpin, a perpetual-calendar watch charm, and a self-pasting scrapbook, the latter being the only invention of his that made any money. He was entranced by the mechanical won-ders he encountered in the semiautomated home of a wealthy railroad magnate, and wrote his wife a giddy thirty-two-page description of such marvels as the house's self-regulating heating systems, rotating dining room table, and water-powered pipe organs. He was fascinated by Tho-mas Edison's inventions and made several phonographic recordings of his voice in Edison's laboratory. And he was intrigued by the electrical experiments he watched his friend Nikola Tesla conduct in his laboratory loft in lower Manhattan. (See photo on p. 177.)

All sorts of innovations excited Twain both in and of themselves and as potential investments: Plasmon, a health food made from dried milk; an envelope maker; an advanced cash register; a process for photograph-ically programming carpet patterns. He would send his long-suffering nephew-by-marriage, Charley Webster, an endless series of instructions to look into this or that promising new invention. "Dear Charley," began a typical missive written in 1884, "Look into that hand-grenade thing & see what the stock can be bought for, & what the condition of the concern is. It is going to do an enormous business some day." Twain may have been the first author to use phonographic dictation in composing one of his books (*The American Claimant*, published in 1892), and he claims to have been the "first person in the world to apply the typing-machine to literature" (portions of *Life on the Mississippi* were submitted to the publisher typed).

Twain saw his first typewriter in 1874 and rushed right out and bought one for $125. His first efforts included a letter to William Dean Howells and one to his brother Orion. To Howells he wrote (all of the letters were capitals):

I DON'T KNOW WHETHER I AM GOING TO MAKE THIS TYPE-WRITING MACHINE GO OR NTO,: THAT LAST WORD WAS INTENDED FOR N-<u>NOT</u>; BUT I GUESS I SHALL MAKE SOME SORT OF SUCCESS OF IT BEFORE I RUN IT VERY LONG. I AM SO THICK-FINGERED THAT I MISS THE KEYS.

YOU NEEDNT A SWER THIS; I AM ONLY PRACTIC-ING TO GET THREE;

<u>ANOTHER SLIP-UP THERE</u>; ONLY PRACTICING TO GET THE HANG OF THE THING. I NOTICE I MISS FIRE & GET IN A GOOD MANY UNNECESSARY LETTERS AND PUNCTUATION MARKS. I AM SIMPLY USING YOU FOR A TARGET TO BANG AT. BLAME MY CATS BUT THIS THING REQUIRES GENIUS IN ORDER TO WORK IT JUST RIGHT.

To Orion he wrote the same day,

I AM TRYING TO GET THE HANG OF THIS NEW F FANGLED WRITING MACHINE BUT AM NOT MAK-ING A SHINING SUCCESS OF IT. HOWEVER THIS IS THE FIRST ATTEMPT I HAVE EVER MADE, & YET I PERCEIVETHAT I SHALL SOON & EASILY ACQUIRE A FINE FACILITY IN ITS USE. I SAW THE THING IN BOSTON THE OTHER DAY & WAS GREATLY TAKEN WITH IT. . . . THE HAVING BEEN A COMPOSITOR IS LIKELY TO BE A GREAT HELP TO ME, SINCE O NE CHIEFLY NEEDS SWIFTNESS IN BANGING THE KEYS. . . .

Twain's analysis of the typewriter's virtues at the close of the letter to Orion suggests that he would have been ecstatic with a laptop:

THE MACHINE HAS SEVERAL VIRTUES. I BELIEVE IT WILL PRINT FASTER THAN I CAN WRITE. ONE MAY LEAN BACK IN HIS CHAIR & WORK IT. IT PILES AN AWFUL STACK OF WORDS ON ONE PAGE. IT DONT MUSS THINGS OR SCATTER INK BLOTS AROUND. OF COURSE IT SAVES PAPER.

Eight years later Twain typed a letter to Karl and Josephine Gerhardt in Paris in which he apologized for typing instead of writing "BECAUSE I AM FULL OF RHEUMATISM AND LAZINESS." But a month later, he had abandoned the idea that any apology was called for, as the figure below demonstrates.

Twain's enthusiasm for technological innovation himself, his widespread recognition as a popular icon, and the continuing interest in his work have made him and his writings appealing vehicles in the design and presentation of a range of new technologies—including models of graphic programming systems, software manuals, computer games, CD-ROMS, and robotics—and in discussions of them in the media. In the October 1989 issue of *PC Magazine*, for example, in "The Marriage of Text and Graphics," by Charles Petzold, two versions of the first paragraph of *Huckleberry Finn* are used to illustrate an extremely technical discussion of the superiority of raster fonts to vector fonts in the display of text on screen, and Gordon McComb begins his book *WordPerfect 5.1: Macros and Templates* (1990) with some thoughts on what Twain would have made of a word processor. Twain serves as a guide on a highly imaginative and visually stunning virtual journey down the Nile in "Nile: Passage to Egypt," a pioneering interactive CD-ROM documentary released in 1995 and developed jointly by the Discovery Channel and the Austin-based multimedia firm Human Code. Another CD-ROM called "Twain's World," which has been out since 1993, allows users who are not deterred by the embarrassingly amateurish graphics to browse through a range of Twain's books, do word searches, and print out favorite passages. And Twain and his works were the subject of the first full-length feature film shot in "Claymation," a patented stop-motion form of animation directed and produced by Will Vinton of dancing California raisin fame. (Sometimes the links between Twain and new technologies prove less than meets the eye: the widely used TWAIN application programming interface standard that facilitates importing images from a scanner

March

DEAR MR. & MRS GERHARDT :-

YOU SEE I AM WRITING THAT
WAY AGAIN. THE FACT IS, I WRITE SO MUCH PLAINER
WITH THE TYPE-WRITER, THAN I CAN WITH THE PEN, THAT
I EXPECT EVERY-BODY TO APPLAUD ME FOR MAKING THE
CHANGE. AS SOON AS YOU GET USED TO THE TYPE-WRITER
YOU WILL BE OFFENDED WHEN PEOPLE WRITE YOU IN ANY
OTHER WAY. I AM TRYING TO FORGET HOW TO WRITE
THE OLD HAND, BECAUSE THE NEW IS SO MUCH PRETTIER.
LET KARL SEND TO MY LONDON PUBLISHERS, CHATTO &
WINDUS, PICCADILLY, LONDON FOR THAT GREEK AND ROMAN
MYTHOLOGY. THEY WILL SEND THE BOOK, AND CHARGE IT TO
ME. KARL CAN SEND THIS LETTER TO THEM AS AUTHORITY,
IF HE IS AFRAID THEY WON'T SEND THE BOOK OTHER-WISE.
OR HE CAN ORDER IT FROM THEM, AND SEND THEM THE
MONEY HIMSELF; THOUGH THEY WILL NOT CHARGE SO MUCH
FOR IT, IF I AM TO DO THE PAYING. IT IS VERY GRAT-
IFYING INDEED, TO HEAR OF KARL'S TRIUMPH AND HIS

18. *Letter to Karl and Josephine Gerhardt typed by Samuel Clemens in 1882. Despite the fact that his typewriter wrote in only capital letters, Twain was delighted with it (Photo courtesy of the James S. Copley Library, La Jolla, California.)*

19. Samuel Clemens participating in an electrical experiment in the New York laboratory of his friend Nicola Tesla. (Photo courtesy of the Mark Twain House, Hartford, Connecticut)

turns out to be an acronym for <u>T</u>echnology <u>W</u>ithout <u>A</u>ny <u>I</u>nteresting <u>N</u>ame.)

In the fields of animatronics and robotics, Mark Twain has had at least three incarnations. The most famous is the life-sized android who has been welcoming visitors to Walt Disney World several times a day since 1982 with his brief tour of American history at the "American Adventure."

After the nation's independence has been won and the westward movement is under way, emcee Twain observes that " 'a whole bunch of folks found out "we the people" didn't yet mean *all* the people,' " whereupon "a Frederick Douglass robot is hoisted up on stage" who then "poles (somewhat improbably) down the Mississippi [and] speaks of the noise of chains and the crack of the whip and of his hope that 'antislavery will unlock the slave prison.' " When visitors to the Dairy Products Building at the Ohio state fair in 1989 got bored with butter sculpture, they could say hello to an animatronic Mark Twain that, according to a reporter on the scene, was "so life-like it's downright eerie":

> The robot is the size of a normal man and looks exactly like Mark Twain. A series of high-tech mechanical components and pneumatic cylinders inside account for its 485-pound weight. When Mark Twain bursts into lively song about his favorite foods, dairy products, you'll swear it's a live person.

Animation and robotics joined forces at the 1993 Virtual Reality Expo in New York City when "Virtual Mark Twain" made his debut. Behind the scenes a human actor—usually McAvoy Layne—picks up what the audience is saying through a hidden microphone. His informal, conversational responses are transmitted to the virtual-reality computer through the motion-sensing equipment he wears, animated by digital puppetry and projected onto a screen. The creator of "Virtual Mark Twain," Gary Jesch, hopes that artificial-intelligence technology might one day allow a "wired" "Virtual Mark Twain" to draw on "a digital library of Mark Twain's works while capturing the essence of his character in complex software."

Mark Twain himself was always intrigued by the new. His interest in science, Paine tells us, "amounted to a passion." As Sherwood Cummings has observed,

> As a citizen of his time of stunning progress, he gloried in the scientific and technical achievements he witnessed. At seventeen he described New York's Croton aqueduct as "the greatest wonder yet," and six years later he joined "the waves of jubilation and astonishment that swept the planet" when the first message was sent via the transatlantic cable. In like manner he greeted other of his century's technological miracles: the telephone, the sewing machine, the electric-powered elevator, the skyscrapers that followed, and in 1907, the Marconi Company's first transatlantic wireless message.

Yet despite his admiration for innovation, Mark Twain remained deeply suspicious as well.

His was no Luddite hostility. Here was a man who embraced new inventions with zest and excitement, who spent—and lost—a fortune backing new technologies. But he recognized that the steamboat, for example, the crowning glory of the machine age of his youth, was both the means through which he realized his fondest dream of becoming a riverboat pilot and the agent of his brother Henry's death. (The steamboat explosion that killed Henry was an accident, but Twain blamed himself for having set in motion the chain of events that left Henry on the boat that blew up). The Paige typesetter, the technology that was to change printing forever, changed nothing but Twain's bank balance. Mark Twain knew firsthand that technology could kill. He knew it could backfire. He knew its realities often failed to keep pace with its promise. Yet he was always eager to try out a new process or invention, to test an improved mechanism for satisfying a need or accomplishing a task. In his ambivalence toward technology he often strikes us as peculiarly modern, a child of our own time, even more than of his.

As early as his first novel, *The Gilded Age,* coauthored with Charles Dudley Warner in 1873, some of this ambivalence comes through. No longer the benign instrument of foreign travel it had been in his previous book, *The Innocents Abroad,* here the steamboat takes on more sinister potential. As Jerry Thomason has observed, in *The Gilded Age*

> the machine becomes a means of death and destruction, personally and economically. When the *Amaranth* blows up, it destroys both human life and life savings. Railroads, which join the steamboats as symbols of progress, also produce as much disappointment as they do hope. Railroad companies become the progenitors of land speculations gone amuck and the political corruption that accompanies it. Later, *Adventures of Huckleberry Finn* (1885) extends this sense of danger produced by machines. As Huck and Jim float naturally down the Mississippi, they are run over by a steamboat, escaping somewhat miraculously with their lives, though the raft is destroyed.

Twain demonstrated fairly early in his career an awareness that technology can impede the very things it is designed to facilitate. In an 1880 sketch about a "telephonic conversation," for example, Twain draws humor from the ways in which the telephone can actually hamper rather than help communication (He does so again in *A Connecticut Yankee,* where the site of Hank Morgan's "Miracle of the Fountain" will be miscommunicated as the "Valley of Hellishness" instead of the "Valley of Holiness.")

It is in *A Connecticut Yankee* that Twain most fully explores the potential of technology as a force of destruction. In the novel's apocalyptic penultimate chapter, the electrical power that Hank marshals to fire his inventive technology becomes a devastating weapon of war. He blows up his factories with dynamite charges, turns eleven thousand charging knights into a wall of corpses through electric fences wired to a dynamo, and fells any knights left alive with Gatling guns. The scientific knowledge Hank had used to improve his world has given him the power to destroy it.

Twain was right to fear the potential for dehumanization and devastation that technology offered; indeed, the unprecedented mass destruction of World War I is eerily prefigured in Twain's fiction, as are some subsequent technological nightmares that became reality in the mid-twentieth century. But coupled with this fear was a concomitant openness to the intriguing liberating possibilities that existed alongside the potential for disaster. Twain never lost his excitement about the changes just around the bend. "The trouble with these beautiful, novel things is that they interfere so with one's arrangements," Twain told the *New York Times* in 1906. "Every time I see or hear a new wonder . . . I have to postpone my death right off."

In his least familiar novel, *The American Claimant,* published in 1892, Mark Twain gives his imagination free rein to concoct hypothetical technologies—none of which is threatening, as the Yankee's technology was, since none of it works. His wild-eyed inventor, Colonel Sellers, who earlier figured in *The Gilded Age*, is brimming over with creative inventions, all of which exist only in his mind or on paper; schemes to materialize the dead, light homes with sewer gas, control the world's climate by moving sunspots, and so on. Yet some of his mad schemes, as Bobbie Ann Mason has observed, curiously prefigure developments that in our day have come to pass, such as fax machines, photocopiers, and DNA cloning. In other works as well, Twain's imagination constructs phenomena that resemble technologies we take for granted today: the "electrophonoscope" we meet in *The Man That Corrupted Hadleyburg and Other Stories and Essays* is, as Cynthia Ozick reminds us, suspiciously like what we know as "television"; and Malcolm Bradbury suggests that in "phrenophones" of "Mental Telegraphy," "the Internet was born." How could a Mark Twain who thought "light-year" to be "without doubt the most stupendous and impressive phrase that exists in any language" fail to be titillated by the vast and instantaneous reach of "cyberspace"?

What would Mark Twain make of the Internet? His affection for Johannes Gutenberg, the inventor of moveable type, might have led him to check out the "Gutenberg Project" and find out what an "e-text" was. As soon as he did, he would be outraged: How could they be giving away so many of his books for free? Twain would demand. And all those other books by other writers? Didn't all his lobbying and consciousness-raising about copyright issues do anything? he'd fume. He thought copyright should last in perpetuity. How on earth could an author make a living if just one copy of his book could reach everyone with an Internet account? Good questions. After his initial ire wore off, he'd surf the web and smile at the idea of finding his books in libraries all over the world and at the thought of reading them in Hindi or Japanese. He might stop a moment to examine a nifty model of Disneyland's *Mark Twain* riverboat constructed solely out of typographical symbols. And he might puzzle over how that unknown prankster's obscene sketch of Uncle Silas in *Huckleberry Finn*—the one that slipped into sales prospectuses before it was caught and corrected—still managed to circulate in cyberspace. He'd marvel at Jim Zwick's phenomenal "Mark Twain Resources on the World Wide Web" at http://web.syr.edu/~fjzwick/twainwww.html, a site that would direct him to some seventy-five to eighty additional Twain sites. And he'd be intrigued by the arcane questions his friends troubled themselves over on the Mark Twain Forum, an e-mail discussion group (What books were published by Mark Twain's Charles Webster Publishing Company besides those written by Twain? Who was Mildred Leo Clemens? Where in Twain's opus could one find the made-up pseudo-German word Mekkamuselmannenmassenmenchenmoerdermohrenmuttermarmormonumentenmacher and what did it mean?) Twain would lurk and maybe float some false factoids when he felt like causing trouble. He wouldn't come up for air for a month. And when he was done, he'd turn off his computer, pull one of his books off the shelf, and fondle it lovingly. He'd smell the binding, turn the pages, feel its heft in his hands. And he'd wonder whether books like this one would someday go the way of steamboat trade on the Mississippi.

EPILOGUE

February 1993. The chill mist that seemed to hang permanently over Cambridge, England, enveloped me as I headed toward the river from my apartment in Clare Hall, the postgraduate college at the University of Cambridge where I was a Visiting Fellow. Crossing Grange Road, I passed the King's College School, where my youngest son was enrolled for the year. When told that his school was founded by Henry VI, he had been unimpressed: the dates of English history were just a blur to him at the time. But his eyes had opened wide when he was told that if you wanted to pass the geography course when the school was set up, you had to know that the world was flat. I crossed the river and entered the central court of a college founded more than five hundred years before the birth of the man I was scheduled to lecture on this evening.

The American Literature Seminar, sponsored by the university's English department, was a small monthly gathering of faculty and graduate students that met in Clare College. I arrived at the seminar room. Introductions were made. One person was working on Henry James, one on Hawthorne, one on T. S. Eliot, one on William Carlos Williams, and so on. Wine was passed around. The atmosphere was convivial as we took our seats at a large round table and I summarized my research on the influence of African-American voices on *Adventures of Huckleberry Finn*. I spoke for an hour and a lively discussion followed for another hour.

Then a silver-haired don who had been completely silent all evening suddenly exploded. "This is going to sound very negative, but I can't figure out what you Americans *see* in this book," he sputtered. "Don't you know it's a *bad* book?"

I could hardly suppress a smile. For Twain had set out to write a book that broke all the rules of English-novel-writing, and the don's outburst was clear testimony to the fact that he had succeeded in doing just that.

"Some Americans actually agree with your appraisal of the novel," I replied. I cited Julius Lester's dismissal of the book as "a fantasy of adolescence," a "dismal portrait of the white male psyche," a bad book in which freedom means simply "freedom from restraint and responsibility." The don looked pleased.

"But," I continued, "you're right that most of us don't share your views." (When I had read Lester's comments aloud to Ralph Ellison in 1991, Ellison was aghast: "Oh my God. Where did he write that? . . . My God—how can he say that it was an irresponsible idea of freedom?") "Twain wanted to do something that hadn't been tried before," I continued. "He wanted to write a book no Englishman could even conceive of at the time—a book written in the voice of a child who skipped school and never learned proper grammar, who was not ashamed of his ignorance and who was convinced—quite rightly—that we would care about what he had to say nonetheless. Everything changed on the literary landscape after this book appeared: it made Hemingway, Faulkner, Ellison—twentieth-century American fiction—*possible*." At this point the graduate students jumped in, and before the evening was over Twain had clearly come out on top.

When the seminar ended, I made my way out of the college gates, turning back for a moment for a last look at the perfectly proportioned, palatial Clare College courtyard. What an appropriate setting, I thought, in which to defend the book which served as America's literary declaration of independence. For Clare had been the college of Charles Townsend, Chancellor of the Exchequer, who had been responsible for imposing the taxes which precipitated what the English call the "American

War of Independence." And Clare had also been the college of Lord Cornwallis, who had surrendered his army to the American troops at Yorktown in 1781. America, young as she is, seems to have had a fairly long history of flummoxing Clare men.

Outside the United States, in other former European colonies, *Huckleberry Finn* is often taught as a model of how one breaks free from the colonizer's culture to create an indigenous national literature. Maria Alejandra Rosarossa, a professor in Buenos Aires, claims Twain as an "American" writer in the hemispheric sense of the word for precisely this reason—his ability to illuminate "a problematic issue" shared by every culture in North and South America: "its colonial origin." She believes he is valued in Argentina in part for his adeptness at showing "the tension between the unauthorized culture, the periphery, the border, and the authorized central one, continuously in conflict, both in the States and in Latin America." Similarly, political scientist Ralph Buultjens recalls that when he was a schoolboy in Sri Lanka, *Huckleberry Finn* was presented as an example of how you create literature in English that has next to nothing to do with England.

*   *   *

The irate Cambridge don would have agreed, most likely, with the British publication that in 1870 found Twain to be "a very offensive specimen of the vulgarest kind of Yankee." This is not to say that all British readers denied Twain respect. On the contrary, Oxford University awarded him an honorary degree in 1907, much to his delight, and some of his staunchest advocates and most sensitive readers both in his time and in ours have been British. One thinks of Andrew Lang, for example, who in 1887 deemed Twain "one of the greatest living geniuses" and chastised cultured critics in the United States for being "not as proud of Mark Twain" as they ought to be. In the twentieth century, Tony Tanner's brilliant readings of Twain come to mind, as do fine recent critical studies by Peter Messent and Peter Stoneley, all of whom appreciate Twain's efforts to write a book unlike any that had gone before. This was, of course, precisely what irked the don. His attitude resonates with the irritation Tom Sawyer expressed toward Huck in *Huckleberry Finn*: "You don't ever seem to want to do anything that's regular," Tom complained, "you want to be starting something fresh all the time." An Englishman, Twain once observed, "is a person who does things because they have been done before." An American, he maintained, is "a person who does things because they haven't been done before."

"What is it that confers the noblest delight?," Twain asked in *The Innocents Abroad*.

> What is that which swells a man's breast with pride above that which any other experience can bring to him? . . . To be the *first*—that is the idea. To do something, say something, see something, before *any body* else—these are the things that confer a pleasure compared with which other pleasures are tame and commonplace, other ecstasies cheap and trivial.

Mark Twain was indeed able to "do something, say something, see something before any body else": he saw the imaginative possibilities the United States offered, and his acts and words and visions changed the literary landscape forever. He disrupted familiar paradigms, smashed stale pieties, and blasted any dogma or system he found in his way. The wildness, the anarchic irreverence, the hilarity and the élan were antidotes to the self-satisfied conventionality, ordered seriousness, and grim purposefulness that Twain knew could be fatal to the development of a "sound heart."

Mark Twain helped his readers extricate themselves from the sensibilities of the Victorian era and begin the march toward modernity. He helped writers narrow the gap between the oral and the written and learn to capture vernacular speech on the page with unprecedented spirit and grace. He modeled how a writer could probe the complexities of contemporary challenges without losing the storyteller's sense of narrative or drama. He did all this in language that was clear, sharp, and meticulously crafted—and he demanded from those around him nothing less.

Mark Twain may not have suspected, when he set off on the *Quaker City* voyage in 1867 that would produce *The Innocents Abroad*, that the journey would mark a milestone in American cultural navigation. The great explorers of the past had planted European flags on the soil of the New World. Twain sailed back the other way and figuratively planted an American flag on the European continent, marking as his own an imaginative terrain that the native inhabitants had for centuries assumed was theirs.

Europe was rich with everything America had none of: castles; relics; royalty; history; ancient legends; certifiably "great" painters, sculptors, composers, and writers. For Europeans, indeed, for most Americans who preceded Twain as travelers to Europe, the proper stance for an American in the face of the grandeur that was Europe was awestruck admiration.

But time and time again Twain refused to genuflect on command. He tells us, for example, that the guide in Genoa

had something which he thought would overcome us. He said: "Ah, genteelmen, you come wis me! I show you beautiful, O, magnificent bust Christopher Colombo!—splendid, grand, magnificent!' . . .

The doctor put up his eye-glass—procured for such occasions:

"Ah, what did you say this gentleman's name was?"

"Christopher Colombo!—great Christopher Colombo!"

"Christopher Colombo—the great Christopher Colombo. Well, what did *he* do?"

"Discover America!—discover America, Oh, ze devil!"

"Discover America. No—that statement will hardly wash. We are just from America ourselves. We heard nothing about it. Christopher Colombo—pleasant name—is—is he dead?"

As Twain debunks both an Old World icon and the sanctity of crediting Europe with the "discovery" of America, the famous "is he dead?" line resonates with the larger trope of European culture as centered on a dead past and American culture as centered on a living present. Twain did not invent the "is he dead?" joke, but he brought it to new heights. "Guides cannot master the subtleties of the American joke," Twain tells us. He celebrated his compatriots' limberness of mind while taking potshots at acolytes of the Old World's ossified artifacts. Like Huck, who "don't take no stock in dead people," Mark Twain and his fellow travelers have limited patience for celebrating dead heroes of European manifest destiny. They are more concerned with capturing new territory across the Atlantic as a province of the American imagination. In *The Innocents Abroad* Mark Twain took on the challenge of making the world safe for an art hewn from the untrammeled, freewheeling vernacular language of his native land.

To clear the way for the success of his own artistic endeavors, Twain had to train his readers to reject the sentimental claptrap that often figured prominently in their notions of "art." For a young man named Grant Wood, who would become one of the best known realist painters of the twentieth century, this strategy provided an epiphany. Wood vividly recalled from his childhood reading of *Huckleberry Finn*

the part wherein Huck staying with the Grangerfords, describes the sentimental pictures a daughter in the family had painted before her premature death. "They was different from any pictures I ever see before—" Huck said, "Blacker, mostly, than is common." And of all the pictures, you will remember the masterpiece was an unfinished study of a grief-stricken young woman ready to leap off a bridge, tears, gown and hair flowing with three pairs of arms—all in different positions. The idea had been to see which pair of arms would look the best and then to scratch out the other two, but the young artist had died before she had made up her mind. "The young woman in the picture," Huck remarked, "had a kind of nice, sweet face but there was so many arms it made her look too spidery, it seemed to me."

As I look back on it now, I realize that my response to this passage was a revelation. Having been born into a world of Victorian standards, I had accepted and admired the ornate, the lugubrious and the excessively sentimental naturally and without questions. And this was my first intimation that there was something ridiculous about sentimentality.

Wood believes his life was changed by his exposure to "the brave way in which—in an age of submission" Mark Twain "lashed out against the artificialities and false standards of his time."

In a similar vein, Bobbie Ann Mason maintains that Twain's "plain style assaulted the wordy romantic rubbish of his day." At a time when "flowery writing"—concentrations of "too many big, unfamiliar and empty words"—had scared many ordinary readers away from "literature," Mason writes, "Twain was one of the first writers in America to deflower literary language." In place of "wordy romantic rubbish" Twain often gives us what Mason calls "rigged-up language," a "language that functions through its potential for inventiveness, just as the necessities of the frontier called for ingenious solutions—making do with what was at hand." Mason, who hails from Kentucky, the state in which Twain's parents were raised, finds this inventiveness deliciously familiar. But precisely what Mason values can strike British purists as beneath respect. As Mordecai Richler reminds us, in March 1995 Prince Charles, speaking at the British Council's English 2000 project,

warned against the threat to "proper English" from the spread of the American vernacular, which he pronounced very corrupting. Because of American influence, he said, "People tend to invent all sorts of nouns and verbs, and make words that shouldn't be. I think we have to be a bit careful, otherwise the whole thing can get rather a mess."

"Obviously Prince Charles has never read Mark Twain . . . or Twain's successors," Richler notes, "and is unaware of how they have enriched a living language that is constantly evolving." He recommends that Prince Charles immediately be sent a copy of *The Innocents Abroad*. Despite the desire of an occasional don or prince to wish it away, Mark Twain's influence on the English language is undeniable and vast. Indeed, Twain is cited in the most recent edition of the *Oxford English Dictionary* more than eighteen hundred times.

If Twain helped clear away the floridly ornate Victorian prose style and invigorated the language with new verbal inventions he also, in *Huckleberry Finn* and other works, ushered into the house of letters a new kind of narrator: someone who spoke in dialect but who was nonetheless authorized to be the central consciousness of the story. Twain dissociated the notion of being good from the capacity to use good grammar. Indeed, in *Huckleberry Finn* the morally base king and duke have some of the "best" speech in the book. Critics have suggested that such memorable twentieth-century narrators as J. D. Salinger's Holden Caulfield in *Catcher in the Rye* or Alice Walker's Celie in *The Color Purple* may be distant relatives of Huckleberry Finn—along with the narrators of Sherwood Anderson's "I Want to Know Why" and Albert Murray's *Train Whistle Guitar*, or Lee Smith's multiple first-person narrators in *Oral History*. Twain's most recent literary legacy in the vernacular narrator department is a fourteen-year-old abused child who is the central consciousness of Russell Banks's gripping 1995 novel *Rule of the Bone*. Bone's observant and sharp first-person narration is slangy, immediate, and engaging. In addition, Bone, like Huck, lacks both self-awareness and self-pity and tries to tell the truth.

Twain's literary legacies reach across national borders to Nigeria, Argentina, Siberia, the Czech Republic, and Japan. They can be found, in fact, wherever writers care about their world and their craft.

In 1991 the Booker Prize was awarded to the novel *The Famished Road* by the London-based Nigerian novelist and poet Ben Okri. "Ben Okri's beautifully written and moving novel combines fantasy and the vision of a child, the supernatural and the here-and-now to convey Nigerian peasant life in a changing world," wrote the chairman of the Booker Prize judges, hailing the book as "one of the most ambitious as well as one of the most fully realized of the year's novels. It brings a distinctively black African way of writing and seeing things into the mainstream of European fiction." After dinner one evening in Trinity College, Cambridge, where he was a Fellow Commoner in Creative Arts the year I was a Visiting Fellow at Clare Hall, I had the chance to ask Okri where he got the idea of telling his story through the eyes of a child narrator. He told me the seed was probably planted by Mark Twain's *Huckleberry Finn*.

The great Argentine writer Jorge Luis Borges claimed Twain as a major inspiration, as did the renowned Russian poet Yevgeny Yevtushenko. Twain also served that role for Czech writer Josef Skvorecky, who was both dumbfounded and entranced when a "pedagogical blunder" introduced him to *Huckleberry Finn* as a child. A number of specific details mystified him: in the "repertoire of European torture," for example, being tarred and feathered and ridden on a rail was missing. Nonetheless, Skvorecky wrote, "the story hypnotized me. I was spellbound by it. I finished it. I re-read it many times. In fact, it was my companion through childhood, adolescence, early adulthood."

While Josef Skvorecky was reading Twain as a child in a small town in Czechoslovakia, Kenzaburo Ōe was reading him at age fourteen in a remote mountain village on the island of Shikoku in Japan. As Ōe's friend and translator John Nathan describes it,

> It seems unlikely that a Japanese schoolboy knowing only the tiny, manageable wilderness of the Japanese countryside could be much moved by Huckleberry's pilgrimage down the vast Mississippi: Ōe was ardently moved. It was Huck's moral courage, literally Hell-bent, that ignited his imagination. For Ōe the single most important moment in the book was always Huck's agonized decision not to send Miss Watson a note informing her of Jim's whereabouts and to go instead to Hell. With that fearsome resolution to turn his back on his times, his society, and even his god, Huckleberry Finn became the model for Ōe's existential hero.

"The heroes of Ōe's fiction," Nathan writes, "are invariably sickened by their experience of 'civilization,' driven on a quest for salvation in the form of personal freedom beyond the borders of safety and acceptance. Brothers to Huckleberry Finn, they are men who have no choice but to 'light out for the territory.' " A "devoted fan of Twain's book," Ōe often paid homage to Twain when, on the occasion of winning the Nobel Prize for Literature in 1994, he was interviewed about the roots of his inspiration as a writer.

In the field of science fiction, Twain gave contemporary writers much more than the time-travel plot which Isaac Asimov credited him with having invented. Consider, for example, his influence on the prolific Frederik Pohl. Reading *Huckleberry Finn* was "a watershed event in my life," Pohl writes. It was the first book he read as a child in which "bad people" ceased to exercise a monopoly on doing "bad things." In *Huckleberry Finn* some "seriously bad things—things like the possession and mistreatment of black slaves, like stealing and lying, even like killing other people in duels—were quite often done by people who not only thought of themselves as exemplarily moral but, by any other standards I knew how to apply, actually *were* admirable citizens." The world Tom and Huck lived in, Pohl writes, "was filled with complexities and contradictions" that resembled "the world I appeared to be living in myself." Mark Twain taught a young reader named Ursula Le Guin important lessons about humor and storytelling. "The first time I read the story about the bluejays trying to fill the cabin with acorns, I nearly died," Le Guin recalls. "I lay on the floor gasping and writhing with joy. Even now I feel a peaceful cheer come over me when I think of that bluejay. And it's all in the way he tells it, as they say. The story is the way the story is told."

If Twain's bluejay yarn in *A Tramp Abroad* captured Le Guin's imagination, *Life on the Mississippi* riveted Willie Morris. Twain "was magic to me as a fledgling writer," Morris tells us, "and still is." It was *A Connecticut Yankee in King Arthur's Court* that tore Arthur Miller away from the football field when he was a teenager. Toni Morrison returned to *Adventures of Huckleberry Finn* when she was honing her own skills as a fiction writer, and returned to it again more recently, finding "genius" in parts of the book that had previously stuck in her mind for their awkwardness. Countless other American writers—including Langston Hughes, William Faulkner, Willa Cather, Saul Bellow, Ernest Hemingway, F. Scott Fitzgerald, Richard Wright, Gertrude Stein, Ralph Ellison, Herman Wouk, Ben Hecht, Henry Miller, William Saroyan, Elizabeth Spencer, and Tillie Olsen—all acknowledge Twain as an important forebear.

Twain helped instill in his fellow writers a respect for clarity and precision in language and an appreciation of what that clarity and precision could yield. For a number of major authors in the twentieth century he became the ultimate writing teacher. In college some time around 1971, for example, when the Black Aesthetic was in ascendance and black artists were being urged to look exclusively to black models, David Bradley decided that he aspired "to be an artist"—perforce a black artist, "for it never occurred to me I *could* be any other kind." While puzzling over the exhortations of apostles of the Black Aesthetic like James T. Stewart (author of "The Development of the Black Revolutionary Artist") and playwright LeRoi Jones, Bradley happened across a book in the stacks of his college library called *Literary Geniuses on Literary Genius* and stopped when he got to a statement by Mark Twain: "The difference between the *almost right* word and the *right* word is really a large matter—'tis the difference between the lightning-bug and the lightning." Bradley was struck by the fact that

> there was nothing *cultural* about the statement. There was nothing political about it. There was nothing middle-class about it, nor anything black or white—except the meaning. . . . If you wanted to construct a literary aesthetic, it seemed like a pretty good place to start.

"Of course," Bradley continues,

> there was a problem: Mark Twain was a white guy. . . . Stewart, Jones, et alia would have rejected him out of hand, and I almost did too. But fortunately [*Literary Geniuses on Literary Genius*] also included a longer piece called "How to Tell a Story," and fortunately I had not grown so narrow-minded as not to read it. That essay made it clear to me that Mark Twain understood the aesthetics of literature a lot better than Stewart or even Jones.

Eventually Bradley found Twain's essays on Cooper and was fascinated to see that Twain's objections to Cooper were more or less what his own had been. And although he "did not at first realize that Twain was being his usual ironic self with all this business about the 'nineteen rules governing literary art in the domain of romantic fiction,' " Bradley writes, "by the time I figured out there was no such list of rules outside Twain's own head, I had decided that the rules made *sense*. It seemed to me they were a pretty good blueprint for writing—negro writing included." But Twain's rules "constituted exactly the kind of 'white model' that Stewart said a black artist had to eschew," Bradley tells us. "Antithetical models" vied for his allegiance; he made a choice.

> I thought about James T. Stewart and all the black artists in the world on the one hand, and Mark Twain on the other. I thought about Jim, the runaway slave, who called himself rich because he had stolen himself, and about Huck, the piece of poor white trash, who first humbled himself to a nigger and then abandoned all hope of Heaven to free that nigger, and about the middle-class white Southern man who created both of them. I thought not only about wanting and needing to write, but about what kind of writer I wanted to be, what kind of

stories I wanted to tell. It seemed to me I had to decide, forever, between two things. I studied a minute, sort of holding my breath, and then I said to myself, "All right," I thought, "I'll *go* to hell."

The result? Bradley's *Chaneysville Incident*, a novel which engages history and memory with a sharpness that Twain himself would have relished, was hailed by one critic as "perhaps the most significant work by a new male black author since James Baldwin dazzled the early '60s with his fine fury," and by another as a book which "may be placed on the honor shelf, right next to Ellison's [*Invisible Man*]."

20. *Author David Bradley (pictured left) chose Mark Twain as his writing teacher. "If we'd eradicated racism in our society," Bradley says, "*Huckleberry Finn *would be the easiest book in the world to teach." (Photo credit: © Carole Patterson)*

21. *Twain's "plain style assaulted the wordy romantic rubbish of his day," writes novelist Bobbie Ann Mason (pictured above). "Twain was one of the first writers in America to deflower literary language." (Photo credit: Thomas Victor, 1985)*

22. *Reading* Huckleberry Finn *was a "watershed event" for "Grand Master" science fiction writer Frederik Pohl, (pictured above), the first book he read as a child in which "bad people" ceased to exercise a monopoly on doing "bad things." (Photo credit: Fred Fox Studios, Ltd.)*

23. *Nobel Prize winning novelist Kenzaburo Ōe (pictured left with his son) was "ardently moved" by Huck's moral courage when he first read Twain's novel as a schoolboy in Japan and Huck became the model for his existential hero. (Photo courtesy of Grove Press, publisher of works by Ōe including* Hiroshima Notes; Nip the Buds, Shoot the Kids; A Personal Matter; *and* Teach us to Outgrow Our Madness.*)*

24. *The judges who awarded London-based Nigerian novelist and poet Ben Okri the Booker Prize in 1991 hailed his book as bringing "a distinctively black African way of writing and seeing things into the mainstream of European fiction." Where did Okri (pictured right) get the idea of telling his story through the eyes of a child? From* Huckleberry Finn. *(Photo credit: © John Foley)*

25. *Nobel Prize winning novelist Toni Morrison, (pictured left) who went back to Twain when she was honing her skills as a writer, believes that "the cyclical attempts to remove the novel from classrooms extend Jim's captivity on into each generation of readers." (Photo credit: Kate Kunz)*

* * *

The controversies that dog Twain's greatest novel show no sign of abating. Its preeminence is regularly challenged in venues ranging from scholarly journals to *Harper's Magazine*. In a 1992 article entitled "Nationalism, Hypercanonization, and *Huckleberry Finn*," published in a highly theoretical journal called *boundary 2*, English professor Jonathan Arac argued that the "hypercanonization" of *Huckleberry Finn* was regrettable. A more suitable candidate for the central place in American letters, he maintained, would be one of the "Leatherstocking" novels of James Fenimore Cooper, unjustly underappreciated. We don't have to guess what Twain's response would have been. Of Cooper's novel *The Deerslayer* Twain wrote,

> There have been daring people in the world who claimed Cooper could write English, but they are all dead now. . . . A work of art? It has no invention; it has no order, system, sequence, or result; it has no life-likeness, no thrill, no stir, no seeming of reality; its characters are confusedly drawn, and by their acts and words they prove that they are not the sort of people the author claims that they are; its humor is pathetic; its pathos is funny; its conversations are—oh! indescribable; its love-scenes odious; its English is a crime against the language.
>
> Counting these out, what is left is Art. I think we must all admit that.

Novelist Jane Smiley hadn't read *Huckleberry Finn* since junior high, but when she broke her leg and was immobilized for a spell, she decided to read it again. She shared her reactions with the readers of *Harper's*

26. *"One of the functions of comedy," said novelist and essayist Ralph Ellison, "is to allow us to deal with the unspeakable. And this Twain did consistently." Here Ellison (left) is pictured with historian C. Vann Woodward. (Photo courtesy of Nancy Lewis and R.W.B. Lewis)*

Magazine in January 1996: "I closed the cover stunned. Yes, stunned. Not, by any means, by the artistry of the book but by the notion that this is the novel all American literature grows out of, that this is a great novel, that this is even a serious novel." The book, she maintains, "has little to offer in the way of greatness." She faults Twain for not taking Jim's aspirations toward freedom seriously: her evidence is that he doesn't let him cross the river to Illinois (she seems unaware of the dangers inherent in that plan in the 1840s) and she faults Twain for demanding too little from Huck: "If Huck feels positive toward Jim, and loves him, and thinks of him as a man, then that's enough. He doesn't actually have to act in accordance with his feelings." (Smiley must have read a different book than I did, because in my edition the crux of the novel involves Huck's decision to "act in accordance with his feelings.") "The very heart of nineteenth-century American experience and literature, the nature and meaning of slavery, is finally what Twain cannot face," Smiley maintains. *Uncle Tom's Cabin*, she argues, is infinitely superior: she'd prefer her children to read Stowe's book and wouldn't care if they never read Twain's. (The possibility that both books might deserve a place in their classroom doesn't seem to cross her mind).

While Jane Smiley keeps Twain's works away from her kids because she considers him soft on racism and without redeeming value, Ralph Wiley, a black writer who specializes in issues of race, is feeding *his* child a rather different fare. While he and his twelve-year-old erupt with laughter at one of Mark Twain's slyly delectable asides, Wiley formulates a theory of what his son needs to get along in life:

> It was at that moment, as my son and I were dying laughing, that I decided I would go forth the next day impelled . . . by my own hypothesis for his survival into the twenty-first century in America: *If you know a little math, can understand Mark Twain's writings and, most difficult of all, can avoid being a victim of ever-ominous Circumstance, you have a fighting chance.*

How could understanding Mark Twain's works be essential to *any* child in twenty-first-century America, let alone a black child? Smiley would shake her head. Not a clue. And she would be totally mystified by Wiley's conviction that "the end of *Huckleberry Finn* is near-perfect." This from the author of *Why Black People Tend to Shout, What Black People Should Do Now*, and *Dark Witness: When Black People Should Be Sacrificed (Again)*.

She would be stumped as well by the passionate defense of *Huckleberry Finn* voiced by a black child in Washington, D.C., who read the book in her sophomore English class in the fall of 1994. Julia Rosenbloom, who is black and Jewish, remembers having been "a little worried about how people would react to the word nigger"; she ended up convinced the book was important for all students to encounter in high school. Twain creates a hierarchy of moral actors in the novel, she maintains, with Tom at the bottom, Huck in the middle, and Jim on top. Twain's irony underscores the fact that the society he describes in the book condemns the person who is most admirable from a moral perspective to the most base condition socially, physically, and legally, she said. The point is that "Twain is ridiculing the society that treats Jim this way." The novel provoked a host of interesting discussions—in class and in the lunchroom and the halls—about racism, about how an avowedly Christian society justified slavery, about what happened in the post-Reconstruction South. As for Twain's use of the offending term "nigger," she and her classmates came to recognize it for what it was: an accurate reflection of the speech that would have existed in the particular time, place, and milieu Twain wanted to evoke. (As Ellison put it, "Mark Twain was being quite realistic when he used the term. If I find it necessary if I'm writing about characters who use that language, I put it down. That's the way it is. We have to learn to come to terms with it.") The book, she believes, has an important lesson to teach: what it really means to behave morally.

The matchless lessons Twain's sharply ironic novels can teach were recognized by New York University Dean Matthew S. Santirocco, who, in consultation with his faculty, decided to issue each incoming freshman at the College of Arts and Science in 1995, along with a map and university catalogue, a copy of *Pudd'nhead Wilson*. Discussions of the book sparked a highly successful yearlong freshman dialogue on identity. (Couldn't they have given them a nice antislavery tract instead, our mystified literalist might demand, or maybe the Emancipation Proclamation?).

Smiley seems to somehow miss the fact that *Huckleberry Finn* came out more than thirty years after Stowe's novel, and that the two books were written to achieve two different ends. One was written to mobilize sentiment against slavery. The other was written to expose the dynamics of racism. In her effort to excoriate the "deeper racism" inherent in Twain's novel, Smiley seems unconcerned that Stowe parcels out brains to her black characters in inverse proportion to the amount of melanin in their skin. She is also obtuse to the growing consensus about the meaning of the last portion of *Huckleberry Finn*.

"Jim is pushed to the side of the narrative" as the novel proceeds, Smiley complains. That is true. Much as "the Negro [had] been pushed into the underground of the American conscience," as Ralph Ellison pointed out many years ago. Rejecting Hemingway's suggestion that readers stop reading the novel before the point at which Jim is taken prisoner, Ellison wrote,

> So thoroughly had the Negro . . . been pushed into the underground of the American conscience that Hemingway missed completely the structural, symbolic, and moral necessity for that part of the plot in which the boys rescue Jim. Yet it is precisely this part that gives the novel its significance. Without it, except as a boy's tale, the novel is meaningless.

More recently, Nobel Prize winner Toni Morrison characterized the last third of the novel as Mark Twain's commentary on the "collapse of civil rights for blacks" in the 1880s, a time when "the nation, as well as Tom Sawyer, was deferring Jim's freedom in agonizing play." As Frederick Douglass put it in his 1880 speech in Elmira, "the old master class" had managed to triumph "over the newly enfranchised citizen." Impressed into chain gangs, "banished from the ballot box and robbed of representation," and forced to "work the farms of their former masters under the lash," "persons of color" throughout the South, Douglass stated, were living in a condition "but little above what it was in the time of slavery." The Supreme Court's 1883 decision to overturn the Civil Rights Act of 1875 rolled the clock back even further when it came to social justice. As he watched these events transpire, Twain finished his novel—not in a slapdash or slopped-together manner, but making subtle and careful revisions.

In the book's famous ending—variously maligned as a failure, a mistake, a retreat, or worse—what do we find? Incarcerated in a tiny shack, with a ludicrous assortment of snakes, rats, and spiders put there by an authority figure who claims to have his best interests at heart, Jim is denied information that he needs and is forced to perform a series of pointless and exhausting tasks. After risking his life to get the freedom that, unbeknownst to him, is already his, after proving himself to be a paragon of moral virtue who towers over everyone around him, this legally free black man is still denied respect—and is still in chains. All of this happens not at the hands of charlatans, the duke and the king, but at the initiative of a respectable Tom Sawyer and churchgoing citizens like the Phelpses and their neighbors. Victor Doyno's research on the manuscript reveals that Twain revised meticulously to make his prose as sharp and effective as possible; his revisions of manuscript page 751, for example, show him emphasizing Jim's bravery and selflessness and contrasting it sharply with Tom's foolishness.

Is what America did to the ex-slaves any less insane than what Tom Sawyer put Jim through in the novel? After all, the spiders and carved grindstones Tom imposed on Jim don't hold a candle to the creativity—or cruelty—of the white cadets at West Point who in 1880 slashed Johnson Whittaker's ears, beat him unconscious, and accused him of staging the attack, alleging he expected to fail his philosophy exam. "It was that one part black that undid him," Twain observed acidly. "It was to blame for the whole thing." Where do we go for a window on the "contrast between our ideals and activities" that was "inescapable" after "the war to 'free the slaves' " as Ralph Ellison put it in our 1991 interview? "People didn't want to talk directly about it," Ellison observed. But Twain did take it on: "One of the functions of comedy," Ellison said, "is to allow us to deal with the unspeakable. And this Twain did consistently." *Huckleberry Finn* may end in farce, but it is not Twain's farce—it is ours. Twain's book is not escapist. It is an escape from the *denial* of the farce we've made of what was—and still is—a noble social and political experiment.

My understanding of history, biography, and literature leads me to read the concluding portion of the novel not as Twain's evasion of responsibility as an artist for what the first part of his novel set in motion (the view advanced by Henry Nash Smith) but as a critique of the United States's evasion of these issues. It is no accident, I believe, that even two of the most astute and far-sighted readers of Twain—Louis J. Budd and Tony Tanner—the first to broach this way of understanding the book, did not articulate this interpretation until the late fifties and early sixties. The consensus that grew gradually from the sixties through the present was built on historians' revisions of nineteenth-century history and its significance and on post–Civil Rights Movement changes in Americans' understanding of the history of American race relations. Writing at a time when pseudoscientific justifications for racism abounded, Mark Twain created a work of art which subverted the reigning social pieties and spoke across time to generations that would continue to struggle with the challenge of extricating their country from the destructive legacies of slavery. I do not claim that this is the only way to read the ending of *Huckleberry Finn*. I also do not wish to argue that any of Twain's contemporaries read it this way, or that Twain consciously had these meanings in mind as he wrote. On the contrary, this reading is shaped indelibly by events of *our* time, by the habits of inquiry and patterns of thought that late-twentieth-century experience has ignited and sustained. Each generation of necessity interprets and understands cultural icons and artifacts with sensibilities shaped by its own time and place. The late-twentieth-century interpretive community has been shaped not only by revisionist historians—both black and white—who rejected the Dunning school of Southern history but also by the voices of distinguished and compelling black writers like Ralph Ellison, Toni Morrison, and David Bradley from whom the reading that follows takes its cue.

In the years that followed the publication of *Huckleberry Finn*, Twain showed himself to be sharply aware of racism's ongoing legacies. Some time in the mid-to-late 1880s he wrote a plot outline for a novel about "passing" set in the North after the Civil War. Twain writes (again prefiguring Malcolm X's remark about what "white racists call black Ph.D.'s"), that the protagonist finds that "even the best educated negro is at a disadvantage, besides always being insulted." Twain never wrote that novel. Nor did he write the nonfiction book on lynchings he at one point planned to assemble. But the serious thought he gave to the possibility of writing the "passing" novel and the lynchings book demonstrates how attuned he remained to the denial of racial justice he saw around him. And his decision against venturing into those minefields makes me appreciate all the more his decision to go the distance in *Huckleberry Finn*. The ending of *Huckleberry Finn*, I maintain, does more than encode and reflect the historical moment in which it was written: with an uncanniness that is as startling as it is sad, it illuminates much of the history that followed it, both in Twain's lifetime and beyond.

What is the history of post-Emancipation race relations in the United States if not a series of maneuvers as cruelly gratuitous as the indignities inflicted on Jim in the final section of *Huckleberry Finn*? Why was the Civil Rights Movement necessary? Why were black Americans forced to go through so much pain and trouble just to secure rights that were supposedly theirs already? You think importing rats and snakes to Jim's shack is crazy? How about this: give blacks the vote, but make sure they can't use it; and when folks call your bluff—like Chaney, Goodman, and Schwerner did—kill them. Or how about this: give blacks an education, but keep from them things that might be useful for them to know about— like the struggle of their ancestors against obscene obstacles and the mean-spiritedness of those who threw those roadblocks in their path (bury the courage of a John Berry Meachum or the callousness of a Judge Pickle and Governor Hobby). And while you're at it, make sure that anything about whites who joined that struggle gets left out as well (forget the

Underground Railroad; shun gadflies like Reverend Tabscott who tell you to remember; bury the magnitude of Twain's own odyssey from being ignorant to being appalled to being actively engaged with the challenge of addressing the whole mind-boggling mess.) Throughout it all, be sure to speechify about the land of the free and the home of the brave and look confused if anyone dares suggest that there might be something wrong with this picture. Wouldn't the Elmira businessmen who refused to give Ernie Davis a summer job or the Hannibal band teacher who couldn't believe a black child could make first chair in the trumpet section invoke not the Fifth Amendment but what Twain called the "silent assertion that nothing is going on which fair and intelligent men are aware of and are engaged by their duty to try to stop"?

Mark Twain got it right—too right: He limned our society's failings with a risky, searing, ironic humor that leaves the reader reeling. Not until Ellison's *Invisible Man* would that dangerous, tricky, borderline out-of-control irony be employed again—to much the same end: that of exposing the "style" with which Tom Sawyer's children have flayed Jim's children raw. How could anything *but* irony address the way American ideals foundered on the fault line of race? Think of the exemplary black high school student at the start of Ellison's novel who is asked to repeat his graduation speech before a gathering of "the town's leading white citizens," a gathering that turns out to be a raucous men's smoker at which he is forced to fight a wild "battle royal" and dive for "gold" coins that give off electric shocks. Spiders? Grindstones? Rats? Snakes? Kids' stuff compared with the trials the folks at the smoker come up with in *Invisible Man*. But in Twain's novel the spiders and grindstone and rats and snakes pain us particularly because of their close proximity to what transpired on that raft, because of the possibility Twain holds out to us that it doesn't have to be that way.

"If the ideal of achieving a true political equality eludes us in reality—as it continues to do," Ellison wrote in his preface to the 1989 edition of *Invisible Man*, "there is still available that fictional vision of an ideal democracy in which the actual combines with the ideal . . . to tell us of transcendent truths and possibilities such as those discovered when Mark Twain set Huck and Jim afloat on the raft." Ellison tells us that Twain's book helped give him the idea that "a novel could be fashioned as a raft of hope, perception and entertainment that might help keep us afloat as we tried to negotiate the snags and whirlpools that mark our nation's vacillating course toward and away from the democratic ideal. . . ." Twain's most famous novel can still be that "raft of hope"—if we let it.

While Americans in the 1880s were attempting "to bury the combustible issues Twain raised in this novel," Toni Morrison writes, Twain insisted on putting them on the table. She believes that "the cyclical attempts to remove the novel from classrooms extend Jim's captivity on into each generation of readers." Letting this book—and the issues it raises—into our minds and into our hearts and classrooms can help us face and address those still explosive and urgent concerns.

"The power of *Uncle Tom's Cabin* is the power of brilliant analysis married to great wisdom of feeling," Smiley writes. Well, "brilliant analysis" is certainly not Huck's strong suit—far from it. ("One has to look at the teller of the tale!" Ellison pleads. It is Huck and not Twain. Why can't readers remember that?) Why does Twain deny Huck any clear-sighted awareness of his society's flaws? Why does he fail to have Huck question the legitimacy of the system that surrounds him even as he acts to sabotage that system? Because the book must end with the reader dissatisfied with any character's take on his world, impatient with the myopia that persists beyond the book's close; it is designed to challenge the reader's complacency and provoke reflection and reconsideration of previously unquestioned assumptions. And that it does—richly—if we let it. That is what makes it such a powerful work of art. Twain's means of getting all this across is a time-honored technique: irony. But I must say of Jane Smiley and those who share her views what Mark Twain said of the townspeople of Dawson's Landing who found themselves so befuddled by lawyer David Wilson's opening remark that they convinced themselves he must be a "pudd'nhead": "irony was not for those people; their mental vision was not focussed for it."

Uncle Tom's Cabin sketches a simple equation of right and wrong and leaves no doubt about what action should be taken. But after the citizens of the United States took the action Stowe was recommending, the country's racial crisis didn't go away: it intensified. The matrix of fraud and hatred and racist betrayals that led to the striking down of the Civil Rights acts in the 1880s, to the upsurge in lynchings and Jim Crow laws in the 1890s, and to the residue of denials and deprivations that exploded in the 1960s and whose repercussions are still with us today is a much more murky topic than the abolition of slavery. There's no happy ending, either. Just struggle. And more struggle. And no end in sight in our time. Faulting Twain for how his story ends is blaming the messenger for the message. "A lot of snotty academics have spent a lot of time and wasted a lot of journal ink criticizing the end of *Huckleberry Finn*," David Bradley observes. "But I notice none of them has been able to suggest, much less write, a better ending. Two actually tried—and failed. They all failed for the same reason that Twain wrote the ending as he did. America has never been able to write a better ending. America has never been able to write any ending at all."

Stowe's book is a bugle call, the prelude to a charge in the right direction (Stowe has no doubt about what direction that is). Twain's book is a wake-up call, an entreaty to rethink, reevaluate, and reformulate the terms by which one defines both personal and national identity, the terms by which one understands a person or a culture as "good" or "evil," a plea to reexamine the hypocrisies we tolerate and the heinous betrayals of hope we perpetrate—in his time and our own—in the name of "business as usual."

* * *

The territory Mark Twain "lit out for" was a strange and complicated place, filled with promise and pitfalls, beauty and barbarity. Twain, like Huck, lit out "ahead of the rest," foreshadowing a host of challenges and conflicts we are still negotiating today. He understood that the best defense against tyranny was laughter, and that the best vaccine against despair was hope. He insisted on taking America seriously. And he insisted on *not* taking America seriously: "I think that there is but a single specialty with us, only one thing that can be called by the wide name 'American,' " he once wrote. "That is the national devotion to ice-water."

Mark Twain understood the nostalgia for a "simpler" past that increased as that past receded—and he saw through that nostalgia to a past that was just as conflicted and complex as the present. He held out to us an invitation to enter that past and learn from it. Are we strong enough to accept?

Twain threw back at us our dreams and our denial of those dreams, our greed, our goodness, our ambition, and our laziness, all rattling around together in that vast echo chamber of our talk—that sharp, spunky American talk that Mark Twain figured out how to write down without robbing it of its energy and immediacy—talk shaped by voices that the official arbiters of "culture" deemed of no importance: voices of children, voices of slaves, voices of servants, voices of ordinary people. Mark Twain listened. And he made us listen. To the stories he told us and to the truths they conveyed. May he continue to goad us, chasten us, delight us, berate us, and cause us to erupt in unrestrained laughter in unexpected places.

NOTES AND SOURCES

Abbreviations

Autobiography
  Mark Twain's Autobiography, ed. Albert Bigelow Paine, 2 vols. (New York: Harper & Brothers, 1924).
CCMT
  The Cambridge Companion to Mark Twain, ed. Forrest G. Robinson (Cambridge and New York: Cambridge University Press, 1995).
Connecticut Yankee
  Mark Twain, *A Connecticut Yankee in King Arthur's Court*, ed. Bernard L. Stein. The Mark Twain Library (Berkeley: University of California Press, 1984).
CTSSE, 1851–1890
  Mark Twain: Collected Tales, Sketches, Speeches & Essays, 1851–1890, ed. Louis J. Budd (New York: Library of America, 1992).
CTSSE, 1891–1910
  Mark Twain: Collected Tales, Sketches, Speeches, & Essays, 1891–1910, ed. Louis J. Budd (New York: Library of America, 1992).
EA
  Elmira Daily Advertiser.
HHT
  Mark Twain's Hannibal, Huck & Tom, ed. Walter Blair. The Mark Twain Papers (Berkeley: University of California Press, 1969)
HT
  J. Hurley and Roberta Hagood, Hannibal, Too: Historical Sketches of Hannibal and Its Neighbor (Marceline, Mo.: Walworth, 1986).
Huckleberry Finn
  Mark Twain, *Adventures of Huckleberry Finn*, ed. Walter Blair and Victor Fischer. The Mark Twain Library (Berkeley: University of California Press, 1985).
HFTS
  Huck Finn and Tom Sawyer Among the Indians and Other Unfinished Stories. ed. Dahlia Armon, Paul Baender, Walter Blair, William M. Gibson, and Franklin R. Rogers. The Mark Twain Library (Berkeley: University of California Press, 1989).
LAT
  Los Angeles Times.
Letters–1
  Mark Twain's Letters, vol. 1, (1853-1866), ed. Edward Marquess Branch, Michael B. Frank, and Kenneth M. Sanderson, with Harriet Elinor Smith, Lin Salamo, and Richard Bucci. The Mark Twain Papers (Berkeley: University of California Press, 1988).
Letters–2
  Mark Twain's Letters, vol. 2 (1867–1868), ed. Harriet Elinor Smith, Richard

Bucci, and Lin Salamo. The Mark Twain Papers (Berkeley: University of California Press, 1990).

Letters–3

Mark Twain's Letters, vol. 3 (1869), ed. Victor Fischer and Michael B. Frank, with Dahlia Armon. The Mark Twain Papers (Berkeley: University of California Press, 1992).

MC&MT

Justin Kaplan, *Mr. Clemens and Mark Twain* (New York: Simon and Schuster, 1966).

MBH

Lorenzo J. Greene, Gary R. Kremer, and Antonio F. Holland, *Missouri's Black Heritage,* (rev. ed.) ed. Gary R. Kremer and Antonio Holland (Columbia: University of Missouri Press, 1980).

MTAYF

Mark Twain at Your Fingertips, ed. Caroline Thomas Harnsberger (New York: Cloud/Beechhurst Press, 1948).

MTE

The Mark Twain Encyclopedia, ed. J. R. LeMaster and James D. Wilson (New York: Garland, 1993).

MTL

Mark Twain's Letters, 2 vols, ed. Albert Bigelow Paine *(New York: Harper Brothers, 1917).*

MTNJ–3

Mark Twain, *Mark Twain's Notebooks and Journals, ed. Frederick Anderson. Vol. III (1883–1891)*, ed. Robert Pack Browning, Michael B. Frank, and Lin Salamo. The Mark Twain Papers (Berkeley: University of California Press, 1979).

NYT

The *New York Times.*

Paine

Albert Bigelow Paine, *Mark Twain: A Biography*, 3 vols. (New York: Harper & Brothers, 1912).

Roughing It

Mark Twain, *Roughing It*, ed. Harriet Elinor Smith and Edgar Maquess Branch, with Lin Salamo and Robert Pack Browning. The Mark Twain Library (Berkeley: University of California Press, 1996).

SCH

Dixon Wecter, *Sam Clemens of Hannibal* (Boston: Houghton Mifflin, 1952).

SLPD

St. Louis Post-Dispatch.

Tom Sawyer

Mark Twain, *The Adventures of Tom Sawyer*, ed. John C. Gerber and Paul Baender. The Mark Twain Library (Berkeley: University of California Press, 1982).

WHB

Shelley Fisher Fishkin, *Was Huck Black? Mark Twain and African-American Voices* (New York: Oxford University Press, 1993).

WP
  The Washington Post.
WTD
  Ron Powers, *White Town Drowsing* (Boston: Atlantic Monthly Press, 1986).

When "OMT" follows the title of a book by Mark Twain, it indicates that the edition being referred to is the first American edition as reprinted in facsimile in *The Oxford Mark Twain* (29 vols.), ed. Shelley Fisher Fishkin. New York: Oxford University Press, 1996.

PROLOGUE

4 *One + signifed* The chart rating the telephone company is reproduced in *Paine*, 2:839.

4 *Write a paper* Anthony Arciola, in Staples High School, Westport, Connecticut.

5 *had been drunk Huckleberry Finn*, 33–34.

5 *had acted very well Huckleberry Finn*, 354.

6 *Do you know what white racists call black Ph.D.'s"?* Malcolm X, *Autobiography of Malcolm X* [1965] With the assistance of Alex Haley. (New York: Ballantine Books, 1973), 284.

7 *all modern American literature* Ernest Hemingway, *Green Hills of Africa* (New York: Charles Scribner's Sons, 1935), 22.

7 *The future historian of America* George Bernard Shaw to Samuel L. Clemens, July 3, 1907, quoted in *Paine*, 3:1398.

7 *In a century we have produced* Twain, autobiographical dictation of Nov. 24, 1906 *Mark Twain in Eruption: Hitherto Unpublished Pages about Men and Events*, ed. Bernard DeVoto (New York: Harper & Brothers, 1940), 376.

7 *By 1906. . . . seventy-two foreign languages.* Statistics come from Robert M. Rodney, ed. and comp. *Mark Twain International: A Bibliography and Interpretation of his Worldwide Popularity* (Westport, Conn.: Greenwood Press, 1982), xxii.

7 *a recent Japanese translation* Hiroshi Okubo, letter to author, July 10, 1995.

8 *Tom Sawyer Bokuju.* Hiroyoshi Ichikawa, letter to author, June 26, 1995.

8 Adventures of Huckleberry Finn *is taught.* For the novel's current presence in classrooms in these locales, I relied on personal communication with the following professors: UNITED KINGDOM: Peter Stoneley, letter to author, June 30, 1995; ARGENTINA: Maria Alejandra Rosarossa, letter to author, Aug. 31, 1995; POLAND: Jerzy Durczak, letter to author, Aug. 4, 1995; CHINA: Wu Bing, letter to author, June 29, 1995/Mei Renyi, letter to author, June 23, 1995; ISRAEL: Barbara Hochman, letter to author, July 4, 1995; GREECE: Katia

Georgoudaki, letter to author, June 29, 1995; BRAZIL: Carlos Daghlian, letter to author, June 29, 1995; INDIA: Prafulla C. Kar, letter to author, June 23, 1995; JAPAN: Makoto Nagawara, letter to author, June 14, 1995; SAUDI ARABIA: A. R. Kutrieh, e-mail to author, July 10, 1995.

8 *Are you an American?* Mark Twain, quoted in Louis J. Budd, "Mark Twain as an American Icon," in *CCMT*, 13.

8 *Against the assault of Laughter* Philip Traum/Satan, "The Chronicle of Young Satan," *Mark Twain's Mysterious Stranger Manuscripts*, ed. William M. Gibson. The Mark Twain Papers (Berkeley: University of California Press, 1969), 166.

9 *the most taught novel* Allen Carey-Webb, "Racism and Huckleberry Finn: Censorship, Dialogue and Change" *English Journal* 82, no. 7 (Nov. 1993): 22.

9 *Hundreds of Twain impersonators* Statistic comes from Louis J. Budd. "Impersonators," *MTE*, 389–91.

10 *In books and articles published over the last ten years* WHB; "Racial Attitudes," *MTE*, 609–15; "Mark Twain and Women," in *CCMT*, 52–73; "False Starts, Fragments and Fumbles: Mark Twain's Unpublished Writing on Race." *Essays in Arts and Sciences* 20 (Oct. 1991): 17–31; "Race and Culture at the Century's End: A Social Context for *Pudd'nhead Wilson*," *Essays in Arts and Sciences* 18 (May 1989): 1–27 (rpt. in *Mark Twain's Humor: Critical Essays*, ed. David E. E. Sloane [New York: Garland, 1993], 359–88); "Mark Twain" in Shelley Fisher Fishkin, *From Fact to Fiction: Journalism and Imaginative Writing in America*, (Baltimore: Johns Hopkins University Press, 1985; rpt. New York: Oxford University Press, 1988), 55–84.

THE MATTER OF HANNIBAL

14 *Hannibal would be the St. Petersburg* The key survey of Twain's refiguring of Hannibal in his fiction remains Henry Nash Smith, "Mark Twain's Images of Hannibal," *Texas Studies in English* 37 (1958): 3–24

15 *the tragic aspects* C. Vann Woodward, *The Burden of Southern History* (New York: Vintage, 1961), 189.

15 *Meachum's river, too* My discussion of John Berry Meachum draws on the following materials: John Berry Meachum, "Preface" to Meachum, *An Address to All the Colored Citizens of the United States* (St. Louis: [n.p.], 1846), 3–6 (pamphlet held in Black Abolitionist Papers Microfilm Collection, document 46.000.00.05:0134); N. Webster Moore, "John Berry Meachum (1789–1854): St. Louis Pioneer, Black Abolitionist, Educator and Preacher," *Bulletin of the Missouri Historical Society* 29 (Jan. 1973): 96–103; Alberta D. Shipley and David O. Shipley, *The History of Black Baptists in Missouri* (n.p.: National Baptist Convention, 1976), 22–23; and *MBH* 67–68. See also Judy Day and M. James Kedro, "Free Blacks in St. Louis: Antebellum Conditions, Emancipation, and the Postwar Era." *Bulletin of the Missouri Historical Society* 30 (Jan. 1974): 117–35; Donnie D. Bellamy, "Free Blacks in Antebellum Missouri, 1820–1860," *Missouri Historical Review* 67 (Jan. 1973): 198–226; idem, "The Education of Blacks in Missouri Prior to 1861," *Journal of Negro History* 59 (Apr. 1974):

143–57. I am grateful to Reverend Robert Tabscott of St. Louis for having shared his knowledge of Meachum with me. He notes that many aspects of Meachum's life remain shrouded in mystery, foremost being how he managed to keep the "freedom school" going within striking distance of hostile authorities. Reverend Tabscott has written a forthcoming children's book on Meachum entitled *Freedom School.*

16 *Bruce Wilson . . . had chosen to investigate an incident* Bruce Wilson, " 'Is It Something I Said?' The Beating of John R. Shillady in Austin, Texas, August 1919," unpublished paper, American Civilization Program, University of Texas at Austin, May 1995.

16 *I told him our Negroes* David J. Pickle, judge in Travis County, quoted in Associated Press report from Austin, Texas, dated Aug. 22, 1919; reprinted in *New York Age*, Aug. 30, 1919, under the headline SHILLADY BEATEN UP BY JUDGE AND CONSTABLE WHILE ON MISSION TO AUSTIN. GOVERNOR APPROVES.

16 *Shillady was the only offender* Governor W. P. Hobby, quoted in "White Negro Advancement Society Men Advised to Stay Out of Texas by Hobby," *Austin American*, Aug. 24, 1919, p. 1. For more on this incident, see John R. Shillady, "Proceedings Planned Against Texas Mobbists," *New York Age*, Sept. 6, 1919, p.1 ; "Austin Beating Sends Shillady on North Trail," AP story, *Austin American*, Aug. 23, 1919; M. M. Johnson, "Only Offender was Shillady," *Austin American*, Aug. 24, 1919; James Weldon Johnson, "Views and Reviews," *New York Age*, Sept. 13, 1919; "Governor Calls Shillady Offender," *New York Age*, Aug. 30, 1919, p. 2.

16 *The county judge and the governor* Travis County judge David J. Pickle was former congressman Jake Pickle's grandfather. Governor W. P. Hobby was former lieutenant governor William P. Hobby Jr.'s father. *The Handbook of Texas*, the key encyclopedia of the history of the state, made no mention of the Shillady incident either in its original edition published in 1952, or in the 1976 supplement. *The New Handbook of Texas* (6 vols. [Austin: Texas State Historical Society, 1996]) does refer to it in the entry on the National Association for the Advancement of Colored People by Michael L. Gillette, (4: 941); the entry on William Pettus Hobby, written by William P. Hobby Jr. (3: 640), makes no mention of the incident. During the years that followed, the former governor was a major media figure in the state, first as president and then as chairman of the board of the *Houston Post.*

17 *We find a piece* The Innocents Abroad, OMT, 165.

19 *John Armstrong traded hay* Advertisement, *Hannibal Journal*, Sept. 21, 1848, cited in *HT*, 103. See also J. Hurley Hagood and Roberta Hagood, "Hannibal's Underground Railroad," *Hannibal Courier-Post*, June 22, 1991, p. 9. In addition to John Armstrong, Hannibal slave traders included, among others, a Thomas Reed, who advertised to buy and sell slaves in every issue of the Hannibal newspaper, and the notorious Beebe, whom Twain mentions by name.

19 *the brutal plantation article* Twain, "Jane Lampton Clemens," posthumously published memoir in *HHT*, 49. See also the following general studies of slavery: *MBH*, 8–62; Harrison A. Trexler, *Slavery in Missouri, 1804–1865* (Baltimore:

Johns Hopkins University Press, 1914); David A. March, "Slavery and Politics," in *The History of Missouri* (New York: Lewis Historical Publishing, 1967), 1: 810–36; Emil Oberholzer, "The Legal Aspects of Slavery in Missouri," *Bulletin of the Missouri Historical Society* 6 (Jan. 1950): 540–45; W. Sherman Savage, "Contest over Slavery Between Illinois and Missouri," *Journal of Negro History* 28 (July 1943): 311–45.

19 *Our feet would crack open* Former slave Emma Knight, 924 North Street, Hannibal, Missouri; Western Historical Manuscripts Collection, University of Missouri, Columbia; rpt. in *The American Slave; A Composite Autobiography. Supplement, series 1. Volume 2: Arkansas, Colorado, Minnesota, Missouri, and Oregon and Washington Narratives*, edited George P. Rawick, Contributions in Afro-American and African Studies, 35 (Westport: Greenwood Publishing, 1977), 202.

20 *was sold on de block* Former slave Clay Smith, 612 Butler Street, Hannibal, Missouri; Western Historical Manuscripts Collection, University of Missouri, Columbia; rpt. in *The American Slave*, 263.

20 *The slave trading at Melpontian Hall* Hagood and Hagood, "Hannibal's Underground Railroad," 9.

20 *Yet the disdain* Twain wrote in "Jane Lampton Clemens," that "the 'nigger trader' was loathed by everybody." *HHT*, 49.

20 *from his home* Twain, "Jane Lampton Clemens," *HHT*, 51. For more on this incident see: *SCH*, 74–75; *Paine*, 1:43; Arthur G. Pettit, *Mark Twain and the South* (Lexington: University Press of Kentucky, 1974), 17–18.

20 *could not have the conscience* Twain, "Jane Lampton Clemens," *HHT*, 51.

21 *I knew the man had a right* *Following the Equator*, OMT, 352.

21 *Everybody seemed indifferent about it* Twain, typescript of notebook 28b, pp. 22–23, Mark Twain Papers; quoted in Pettit, *Mark Twain and the South*, 15.

21 *Everybody granted* *The Tragedy of Pudd'nhead Wilson*, OMT, 303.

21 *as a rule our slaves* Twain, "Jane Lampton Clemens," *HHT*, 49.

21 *Mistress always told us* Emma Knight, quoted in *The American Slave*, 203.

21 *Father run away.* Clay Smith, quoted in *The American Slave*, 263.

21 *loud and frequent groans* Paine, 1:17.

21 *In 1847, when Twain was eleven* Paine, 1:63–64; *SCH*, 148. The *Hannibal Journal* of Aug. 19, 1847, reported: "While some of our citizens were fishing a few days since on the Sny Island, they discovered in what is called Bird Slough the body of a negro man. On examination of the body, they found it to answer the description of a negro recently advertised in handbills as a runaway from Neriam Todd, of Howard County. He had on a brown jeans frock coat, home-made linen pants, and a new pair of lined and bound shoes. The body when discovered was much mutilated." Quoted in *SCH*, 148.

22 *Indeed, slaves escaped* HT, 104.

22 *impressive pauses* Samuel Clemens to Joel Chandler Harris, Aug. 10, 1881; in *MTL*, 1: 402–3.

22 *a gay and impudent and satirical young black man* Twain, "Corn-Pone Opinions" [1901], *Europe and Elsewhere*, ed. Albert Bigelow Paine (New York: Harper, 1923), 399.

22 *a dozen black men and women* Twain, "Jane Lampton Clemens," *HHT*, 49.

22 *Even the children . . ."* *A Connecticut Yankee*, 198.

23 *In those old slave-holding days* Twain, comment in notebook 28, quoted in Walter Blair, *Mark Twain and "Huck Finn"* (Berkeley: University of California Press, 1960), 144.

23 *What is a "real" civilization?* Twain, "On Foreign Critics," speech delivered in Boston, Apr. 27, 1890; in *CTSSE, 1851–1890*, 942.

23 *sound heart* In a notebook entry Twain characterized Huck's dilemma like this: "In a crucial moral emergency . . . a sound heart & a deformed conscience come into collision & conscience suffers defeat." Quoted in Walter Blair and Victor Fischer, "Foreword," *Huckleberry Finn*, xx.

24 *no one knows for sure* WTD, 24–25. The following account of the legend is from *WTD*, 25.

25 *Starting in the 1870s* See J. Hurley Hagood and Roberta (Roland) Hagood, "Log Rafts Bring Prosperity to Hannibal," in Hagood and Hagood, *Hannibal Yesterdays: Historic Stories of Events, People, Landmarks and Happenings in and near Hannibal.* (Marceline, Mo.: Jostens, 1992), 121–26; *WTD*, 26.

25 *the town which furnished the background HT*, 285.

25 *Mark Twain's Boyhood Home HT*, 290.

25 *350,000 visitors . . . thirteen million dollars* Statistics from "Hannibal Tourism Facts and Figures." (n.d., received June 1995), courtesy of Faye Bleigh, director, Hannibal Visitors and Convention Bureau.

25 *Missouri claims Mark Twain for its very own* Walter Williams, introduction to "Mark Twain," in *Five Famous Missourians*, by Wilfred R. Hollister and Harry Norman (Kansas City, Mo.: Hudson-Kimberly, 1900), 7.

26 *Mark Twain's First One Hundred Years HT*, 288–90. See also J. Hurley Hagood and Roberta (Roland) Hagood, *The Story of Hannibal* (Hannibal: Standard Printing Co., 1976), 177–202.

26 *At the Sesquicentennial celebrations in 1985* For a lively book-length account of the sesquicentennial see *WTD*. See also *HT*, 285–300.

28 *Not a sound, anywheres Huckleberry Finn*, 156.

28 *the nice breeze springs up. Huckleberry Finn*, 157.

28 *the white town drowsing Life on the Mississippi*, OMT, 63.

28 *the streets empty or pretty nearly so Life on the Mississippi*, OMT, 63.

28 *the great Mississippi Life on the Mississippi*, OMT, 63.

28 *the dense forest away on the other side Life on the Mississippi*, OMT, 64.

28 *a negro drayman Life on the Mississippi*, OMT, 63.

29 *exertions of good-humored "JOHN"* "Fire," *Hannibal Missouri Courier*, May 15, 1851; quoted in "Biographical Directory," 322, *HFTS*, 321–22.

29 *John Hanicks' laugh MTNJ-3*, 355.

29 *giving his "experience" MTNJ-3*, 355.

29 *John Hannicks, with the laugh* Twain, "Villagers of 1840–3 ," *HFTS*, 102.

29 *In the "South Connecticut Yankee,* 346.

29 *to enslave them by another name MBH,* 64. See also Lawrence I. Berkove, "Free Man of Color," in *MTE,* 301–2.

30 *The flood protection* Henry Sweets, taped interview by author, Hannibal, Mo. (Mark Twain Boyhood Home and Museum), June 21, 1995. All of Henry Sweets's comments quoted here were made during this interview.

32 *In the book section, I've tried to have all of his books* Martha Adrian, taped interview by author, Hannibal, Mo. (Mark Twain Book and Gift Shop), June 21, 1995. All of Martha Adrian's comments quoted here were made during this interview.

33 *To be as nice and as pleasant and friendly.* Faye Bleigh, taped interview by author, Hannibal, Mo. (Hannibal Visitors and Convention Bureau), June 21, 1995. All of Faye Bleigh's comments quoted here were made during this interview.

35 *covered the Jim Crow legislation* Mayor Richard Schwartz, taped interview by author, Hannibal, Mo. (Mayor's Office, City Hall), June 21, 1995. All of Mayor Schwartz's comments quoted here were made during this interview.

37 BAPTISTS APOLOGIZE FOR SIN OF SLAVERY Patricia Rice, "Baptists Apologize For Sin Of Slavery; Blacks And Whites Clasp Hands, Vow Reconciliation." *SLPD,* June 21, 1995, p. 1A.

38 *I can't repent for him* Richard Land, president of the Southern Baptist Convention's Christian Life Commission, quoted in Bill Clough, "Southern Baptists Approve Racial Reconciliation," United Press International, June 20, 1995.

38 *My ancestors had slaves and I'm very proud of it* Unnamed Southern deacon, quoted by Convention leader Rev. Jere Allen in Andrea Stone, "Southern Baptists 'Repent' Past/Apology for Pro-Slavery Stance Due" *USA Today,* June 20, 1995, p. 3A.

38 *discredit to those great men who founded this convention* Cary Kimbrell, quoted in David Waters, "Baptists Repent of Racism, Slavery," *The Commercial Appeal* (Memphis, Tenn.), June 21, 1995, p. 1A.

38 *was an evil . . . from which we continue to reap* "Text of Resolution," *The Commercial Appeal* (Memphis), June 21, 1995, p. 6A.

39 *This delightful two-hour pageant Hannibal & Mark Twain Lake 1995 Visitor's Guide* (Hannibal: Hannibal Visitors and Convention Bureau, 1995), 12.

40 *The slavery question* Frederick Jackson Turner, "The Significance of the Frontier in American History." Paper presented at the meeting of the American Historical Association, Chicago, July 12, 1893. Reprinted in Frederick Jackson Turner, *The Frontier in American History* (New York: Henry Holt, 1953), 24.

40 *"Freedom School"* An alumnus of Meachum's "Freedom School," James Milton Turner, worked effectively to increase educational opportunities for blacks in Missouri after the Civil War, and served as the first black in the U.S. diplomatic corps. For more on his career, see Gary R. Kremer, *James Milton Turner and the Promise of America: The Public Life of a Post–Civil War Black Leader* (Columbia: University of Missouri Press, 1991).

40 *My son wasn't allowed* Ruth Baker, interview by author, Hannibal, Mo. (Mark Twain Family Restaurant), June 22, 1995. All of Ruth Baker's comments quoted here were made during this interview.

41 *Right here* Estel Griggsby, interview by author, Hannibal, Mo. (Mark Twain Family Restaurant), June 22, 1995. All of Estel Griggsby's comments quoted here were made during this interview.

42 *The Cave at Hannibal* Article quoted in Charles A. Norton, *Writing Tom Sawyer; The Adventures of a Classic* (Jefferson, N.C.: McFarland & Co. 1983), 10–11.

42 *an invitation to plot a story* Norton, *Writing Tom Sawyer*, 11.

42 "SAVAGE" INDIAN DISCOUNTS STORY I was not able to identify the newspaper in which the unsigned, undated clipping in the glass frame had appeared.

43 *Joe Douglas was black.* Hannibal historians Hurley and Roberta Hagood confirmed my inference when I asked them about it the following day. In *Hannibal Yesterdays* the Hagoods note that "the Douglasville community had been named for Joe Douglas," who had "acquired property in the area by saving his meager earnings" (74). Douglas was buried by the African Methodist Episcopal Church (gravestone inscription, Mt. Olivet Cemetery, Hannibal).

44 *You don't need to bait your hook* Anon., "Mark Twain Is Still Hannibal's Honored. . . ." [portion of article destroyed], *Hannibal Morning Journal*, May 31, 1902, pp. 1, 5. On glass-covered display boards at the Mark Twain Boyhood Home and Museum.

44 *to meet the exigencies of romantic literature* Twain [autobiographical dictation, Thursday, Mar. 8, 1906], *Autobiography*, 1:175.

44 *another source says he was raised by a black family* Eighty-four-year-old Hannibal resident Clarence Shaffer, quoted in Paul Hendrickson, "Tom Sawyer Days in the Land of Twain; As American as . . . Tom Sawyer Days in Hannibal, Mo.; Tricycles, TV and the Tonic of Tradition in Hannibal." *WP*, July 5, 1978, p. B1. Shaffer produced a copy of Joe Douglas's death certificate for an unnamed "observer" on whose notes the *Post* reporter based his story.

45 *Joe Douglas, good-humoredly called "Injun Joe"* "Injun Joe's Alibi," undated clipping from newspaper entitled *Grit* (presumed date of 1894 comes from known birth date of Joe Douglas and age stated in article) rpt. in *The Twainian*, 10, no. 5 (Sept.–Oct. 1951):4, under the title "The Characters Mark Twain Used in His Writings."

45 *once tried to reform Autobiography*, 2:175.

45 *drunkard and murderer SCH*, 151.

46 *real Injun Joe SCH*, 151.

46 *In 1870, the population of Hannibal* Hagood and Hagood, *Hannibal Yesterdays*, 73–74 (Some time later the school was renamed for abolitionist Frederick Douglass. But as late as 1995 local black resident Ruth Baker, who had attended the school, would still recall the school as having gotten its name from a Hannibal man.)

47 *Hurley Hagood told me the story* Hagood, taped interview by author, Hannibal, Mo. (Mt. Olivet Cemetery), June 23, 1995.

48 *His house in Douglasville was there until 1994.* Hagood, taped interview by author, Hannibal, Mo. (driving through Hannibal), June 23, 1995.

48 *"Mrs. Lula Clay?* Dixie M. Forte, Hiawatha Crow and Rev. Anne Facen, taped interview by author, Hannibal, Mo. (Mark Twain Family Restaurant), June 23, 1995. (Madam C. J. Walker [1867–1919], the first black woman millionaire in America, founded a company that sold hair-care products for African-Americans through field representatives like Mrs. Lula Clay.)

51 *Black John and Tom Blankenship were naturally leading spirits* John Ayres, "Recollections of Hannibal," *Palmyra Spectator*, Aug. 22, 1917, in Morris Anderson scrapbook, photocopy in Mark Twain Papers, *WHB*, 28–29.

51 *the young black man named Jerry, the Missouri slave* See *WHB*, 53–76.

53 *schoolboy days Autobiography*, 1:101.

54 *an abolitionist from his boyhood Autobiography* 2:272.

54 *Go straight ahead* Hurley Hagood, taped interview by author, Hannibal, Mo. (driving through Hannibal), June 23, 1995. The best sources on the Underground Railroad in Hannibal are the chapter on the subject in *HT* and Hagood and Hagood, "Hannibal's Underground Railroad," *Hannibal Courier-Post*, June 22, 1991, p. 9.

54 *In that day for a man to speak out openly* Twain, "A Scrap of Curious History," in *What Is Man?* (1917 ed.); rpt. in *The Complete Essays of Mark Twain*, ed. Charles Neider (Garden City, N.Y.: Doubleday, 1963), 517–23.

54 *clever, witty* Orion Clemens in the *Daily Journal*, quoted in *HT*, 110.

55 *a woman and her son hoeing tobacco* Alanson Work, letter from Palmyra jail to "a brother in Quincy," reproduced in full in George Thompson, *Prison Life and Reflections; or, A Narrative of the Arrest, Trial, Conviction, Imprisonment, Treatment, Observations, Reflections, and Deliverance of Work, Burr and Thompson, who suffered an Unjust and Cruel Imprisonment in Missouri Penitentiary, for Attempting to Aid Some Slaves to Liberty, Three Parts in One Volume.* (Hartford: A. Work, 1851), 17. The events leading up to the trial and the trial itself are also described in R. I. Holcombe, *History of Marion County, Missouri* [1884]; rpt. Marion County Historical Society (Hannibal, Mo., 1979), 239, 263–66. See also Trexler, *Slavery in Missouri*, 121–22.

55 *found that she wanted to be free* Thompson, *Prison Life*, 17.

55 *started in a footpath for the river* Thompson, *Prison Life*, 17.

55 *my negroes* Testimony of William P. Brown under oath, quoted in "Bill of Exceptions," "State vs. Burr, Work and Thompson, Indictment for Larceny," *Narrative of Facts, Respecting Alanson Work, Jas. E. Burr, & Geo. Thompson, Prisoners in the Missouri Penitentiary, For The Alleged Crime of Negro Stealing.* Prepared by a Committee. [For the Quincy, Ill., Anti-Slavery Concert for Prayer, 1842] (Quincy, Ill.: Printed at the Quincy Whig Office, 1842), p. 32 (Microfiche 3463, 1970; Perry-Castañeda Library, University of Texas, Austin)

55 *lamented that Missouri had no "Lunatic Asylum"* The Daily Evening Gazette, Sept. 16, 1841; quoted in Trexler, *Slavery in Missouri*, 122.

55 *a tremendous excitement all over the country* George Thompson, *Prison Life*, 48.

56 *The slaves own themselves* George Thompson, letter from Palmyra Jail, September 4, 1841; rpt. *Narrative of Facts*, 17.

56 *I owns mysef Huckleberry Finn*, 57.

56 *was a fool* Twain, "Villagers of 1840–3," *HFTS*, 102

56 *hearts were so filled* George Thompson, *Prison Life*, 76.

56 *Guilty and Twelve Years* Thompson, *Prison Life*, 79. The men ended up being pardoned after serving the following time: Alanson Work: 3 years, 6 months, 7 days; James Burr: 4 years, 6 months, 17 days; George Thompson: 4 years, 11 months, 12 days. Thompson, *Prison Life*, 273.

57 *he was said to have helped some four thousand slaves reach freedom* "While in Illinois and Missouri, [Alanson Work] helped nearly 4,000 slaves to reach freedom by means of the 'Underground Railroad.' " "Henry Clay Work" [biographical sketch], in *Songs of Henry Clay Work*, comp. Bertram G. Work (New York: J. J. Little & Ives, (c. 1884), rpt. Henry Clay Work, *Songs*, introd. by H. Wiley Hitchcock. Early American Music, 19 (New York: Da Capo Press, 1974), 5. "In Illinois and Missouri [Alanson Work] aided about four thousand slaves to escape by maintaining his home as one of the "stations" of the Underground Railroad." *Dictionary of American Biography*, vol. 10, ed. Dumas Malone (New York: Charles Scribner's Sons, 1936), 531–32. The pardon Work received from the governor was also a banishment that required him to leave Missouri and join his family in Connecticut.

57 *Our country has produced few song writers.* Anon., "Henry Clay Work," *Hartford Daily Courant*, June 9, 1884.

58 *the historian's book* R. I. Holcombe, *History of Marion County, Missouri* [1884]; rpt. Marion County Historical Society (Hannibal, Mo.: Walsworth, 1979).

58 *The author of "Marching Through Georgia"* Karl Gerhardt to Samuel Clemens, Sept. 22, 1887, quoted in note 102, *MTNJ-3*, 335. Mark Twain had mixed responses to Work's songs. In 1866 he expressed his impatience with the overexposure "Marching Through Georgia" had received. "I wish Sherman had marched through Alabama," he wrote (*Mark Twain's Notebooks and Journals* [1855–1873], ed. Frederick Anderson, Michael Frank and Kenneth Sauderson. The Mark Twain Papers [Berkeley: University of California Press, 1975]. I: 228). On the other hand, in 1879 Twain wrote William Dean Howells that at the Army of Tennessee Banquet for General Ulysses S. Grant that Twain had recently attended "somebody struck up 'When we were Marching through Georgia.' Well, you should have heard the thousand voices lift that chorus & seen the tears stream down. If I live a hundred years I shan't ever forget these things—nor be able to talk about them." Clemens to Howells, Nov. 17, 1879, in *Mark Twain–Howells Letters: The Correspondence of Samuel L. Clemens and William D. Howells, 1872–1910*, ed. Henry Nash Smith and William M. Gibson with Frederick Anderson, 2 vols. (Cambridge, Mass.: Harvard University Press, 1960), 1:280.

58 *I tried to make it out to myself Huckleberry Finn*, 124.

58 *got aboard the raft feeling bad and low Huckleberry Finn*, 127.

58 *It would get all around Huckleberry Finn*, 268–69.

59 *your runaway nigger Huckleberry Finn*, 269.

59 *good and all washed clean.* Twain, *Huckleberry Finn*, 269.

59 *got to thinking Huckleberry Finn*, 270–71.

59 *on the expert way in which the whip was handled A Connecticut Yankee*, 199.

60 *It is commonly believed that an infallible effect* Twain, "Jane Lampton Clemens," *HHT*, 51.

60 *bad faith* Forrest G. Robinson, *In Bad Faith: The Dynamics of Deception in Mark Twain's America* (Cambridge, Mass.: Harvard University Press, 1986).

60 *lie of silent assertion* Twain, "My First Lie and How I Got Out of It," in *CTSSE, 1891–1910*, 440.

60 *It would not be possible . . . for a humane and intelligent person* Twain, "My First Lie and How I Got Out of It," *CTSSE, 1891–1910*, 440.

60 *silent assertion that nothing is going on* Twain, "My First Lie and How I Got Out of It," *CTSSE, 1891–1910*, 441.

61 *It was a problem the whole country shared.* Writing in 1989, James O. Horton and Spencer R. Crew stated, "In the 1980s the contributions and experiences of black people are still often excluded from the public presentation of our nation's history. . . . Many museums have minimized the experience of Afro-Americans and their centrality to American history. Too often the burgeoning scholarship in black history has not found its way into the public presentations that have shaped the historical consciousness of millions of Americans." In *History Museums in the United States: A Critical Assessment*, ed. Warren Leon and Roy Rosenzweig (Urbana: University of Illinois Press, 1989), 215–16. In their contribution to this volume ("Afro-Americans and Museums: Toward a Policy"), Horton and Crew trace this gap between "recent scholarship in Afro-American and social history and the public exhibition of American history" to "institutional unwillingness to recognize the importance of Afro-American history to a realistic understanding of American culture" (216). They largely base their analysis of "institutional responses to recent Afro-American historical scholarship" on empirical data gleaned from a survey of 104 museums. They observe that in addition to serving as a clearinghouse for the "now well over one hundred" museums around the country "focused on black history and culture," the African-American Museum Association in Wilberforce, Ohio, which was founded in 1978, has "increased pressure on traditional museums and historic sites to improve their representation of Afro-Americans in their exhibitions and public presentations. To their credit, several museums accepted the challenge, seeking new ways to include black history" (224).

62 *a project that would commemorate Underground Railroad sites* Conversation with James O. Horton at the Dept. of Interior, Washington, D.C., February 8, 1995, and personal communication, May 7, 1996. The National Park Service completed its study in the fall of 1995 in response to a request of Congress. The interdisciplinary Advisory Committee, chaired by historian Charles L. Blockson of Temple University, recommended "consideration of five options, including the development of historic trails, preservation of important Underground Railroad sites like the Parker House in Ripley, Ohio, and the establishment of either several regional or one national interpretive center."

"The National Park Service Completes Study." *Friend of a Friend: The Newsletter of the National Underground Railroad Freedom Center*, 1, no. 1 (Apr. 1996):5.

62 *underground railroad museum being planned in Cincinnati* Funds are currently being raised for the National Underground Railroad Freedom Center in Cincinnati, under the direction of Edwin J. Rigaud, vice president of government relations for the Procter & Gamble Company, on loan to the National Underground Railroad Freedom Center as its first full-time director. The Freedom Center, which is projected to open in 2000 or 2001, "will use emerging communications technology such as the internet and satellite video to link hundreds of historic sites and organizations with both the national center in Cincinnati and with each other." Anon., "Consultant Team Soon to Complete Feasibility Study," "First Director," "Why Cincinnati?" in *Friend of a Friend*.

62 *The experience of slavery and the institutional racism* John Vlach, interview by Charlayne Hunter-Gault in Washington, D.C. at the Martin Luther King Jr. branch of the Washington, D.C., Public Library, broadcast on the *MacNeil/Lehrer NewsHour*, Feb. 5, 1996. Also Vlach, interview by the author, Austin, Texas (LBJ School), Feb. 24, 1996.

62 *See this famous river town* Hannibal & Mark Twain Lake *1995 Visitor's Guide*, 7.

64 *Two things you'd worry about.* Tour guide's comments over loudspeaker, taped during 4 p.m. cruise on the *Mark Twain* riverboat in Hannibal, June 23, 1995.

64 *The island coming up on our right* Tour guide's comments.

000 *the Tom Sawyer of the political world* Twain, *Mark Twain in Eruption*, 49.

66 RACIST YEARBOOK INCIDENT. Jacques Steinberg, "Racist Yearbook Incident Reveals Rift in Greenwich." Boxed quote: "Reactions to a racial incident range from anger to 'they're kids.' " *NYT* national ed., June 21, 1995, p. B4.

66 NIGGER CHECK POINT Photograph of sign accompanied story by Kevin Johnson, "Racist rally reviewed today. 'Roundup' in Tenn. erodes public's faith." *USA Today* July 21, 1995, p. 5A. Photo credit: WJLA-TV via AP. Caption "Good Ol' Boys Roundup. Video images of a racist poster at a 1990 gathering of law-enforcement officers in Tennessee." The article notes that "reports and videotapes [show that] the annual 'Good Ol' Boys Roundup,'— coordinated by Gene Rightmyer, a former agent with the Treasury's Bureau of Alcohol, Tobacco and Firearms—featured racist slogans, offensive entertainment and the sale of T-shirts with racist themes." See also Tim Weiner, "F.B.I. Says at Least 7 Agents Attended Gatherings Displaying Racist Paraphernalia, *NYT*, July 19, 1995, p. A12. Allegations that the videotape had been tampered with were dismissed; the tape was analyzed by the F.B.I.'s Jerry Seper, " 'Good Ol' Boys Roundup' Video Genuine: FBI Finds No Editing in Racist Scenes." *The Washington Times*, Dec. 7, 1995, p. A4.

67 *This is Meachum* Rev. Robert Tabscott, taped interview by author, St. Louis, Mo. (office of the Elijah Lovejoy Society), June 24, 1995, and telephone conversation, June 21, 1995.

68 *When are you going to decide* Robert Tabscott, "Confessions of a White Man," commentary, *SLPD*, Nov. 24, 1993, p. 7B.

68 *free Negroes . . . not born in Missouri* Robert Tabscott, "A Long History of Racial Intolerance," commentary, *SLPD*, Jan. 17, 1994, p. 7C.

68 *received a telephone death threat* Rev. Tabscott noted, "This was not my first encounter with the Klan nor the first threat on my life. I met the wizards of the night 30 years ago in Virginia when a cross was burned in our churchyard. Then again in Jackson, Miss. As the decade of the '60s wore on and a generation of us took our stand against racism in the South. I used to take my children to and from school for safety. I was admonished by the church elders to keep my place or to face expulsion from my work. I was expelled. The creed of the Klan was shared by many of my sophisticated Presbyterian congregation." A Long History of Racial Intolerance," commentary, *SLPD*, Jan. 17, 1994, p. 7C.

68 *People are concerned* Robert Tabscott, "A Long History of Racial Intolerance," commentary, *SLPD*, Jan. 17, 1994, p. 7C.

69 *It's this extravagant, intoxicating ability* Tabscott interview.

EXCAVATIONS

72 *About that time a colored cadet* William Dean Howells, *My Mark Twain* (New York: Harper and Brothers, 1910), 35–36. By the fall of 1995, I became convinced that the cadet to whom Twain referred had been Whittaker and called Prof. John Marszalek, whose book on Whittaker had set in motion the posthumous commission. Philip Butcher had made the same inference in an article in 1969 ("Mark Twain's Installment on the National Debt," *Southern Literary Journal* 1 (Spring, 1969): 48–55). Prof. Marszalek agreed. The most in-depth treatment of the case is John F. Marszalek, *Assault at West Point: The Court-Martial of Johnson Whittaker* (New York: Collier Books, 1994; originally published as *Court-Martial* [New York: Scribners, 1972]). A "Showtime" channel movie, *Assault at West Point*, which was first broadcast in 1994, dramatized Marszalek's research.

72 *that which was base* Twain [Samuel Clemens], autograph manuscript of *Pudd'nhead Wilson*, Pierpont Morgan Library, New York, MA 241–42; quoted in *WHB*, 121–22. This passage was one of many that did not make it into the published version of the book.

72 *Try never to injure another* John Harris, "The Late Lieutenant: Black Cadet Finally Gets Commission Denied in 1880," *WP*, July 25, 1995, p. E1.

72 *like we do hogs down South* "115 Years Later, West Point Commissions Wronged Cadet," *Chicago Tribune*, July 19, 1995, p. 2.

72 *"inferior" and "superior"*; "Negroes are noted . . . Oh, it's just like . . . "Seeking 'Fair Deal' for a Black Cadet." *NYT*, Jan. 31, 1994, p. A10.

73 *burned all the papers* "Senator wants to right an 1870s wrong; Army commission sought for beaten, discharged black cadet," *Dallas Morning News*, Feb. 13, 1994, p. 1A.

73 *a man who through courage* "Black Cadet Receives Posthumous Honor," *The Record* (Bergen, N.J.), Aug. 25, 1995, p. A10.

73 *rectify a grievous injustice . . . a message that injustice and discrimination* Steve Piacente, "Book, movie brought wrong to light." *The Post and Courier* (Charleston, S.C.), July 20, 1995, p. A15.

73 *the moral situation of the United States* Ralph Ellison, taped interview with the author, New York City (Ellison's apartment), July 16, 1991.

74 *surrounded with chains* Samuel Langhorne Clemens to Jane Lampton Clemens, Aug. 24, 1853, *Letters–1*, 4.

74 *in these Eastern States niggers* Samuel Langhorne Clemens to Jane Lampton Clemens, Aug. 24, 1853, *Letters–1*, 4.

75 *Haven't you a friend in the world?* Jervis Langdon, quoted in *Autobiography*, 1:110–11.

75 *Father.* Twain's affection and respect for his father-in-law are clear from the eulogy that he published in the *Express* two days after his death: "Mr. Langdon was a great and noble man, in the best and truest acceptation of those terms. He stood always ready to help whoever needed help—wisely with advice, healthfully with cheer and encouragement, and lavishly with money. He spent more than one fortune in aiding struggling unfortunates in various ways. . . ." Clipping in Mark Twain Papers; quoted in Jeffrey Steinbrink, *Getting to Be Mark Twain* (Berkeley: University of California Press, 1991), 126–27.

78 *Next to possessing genius one's self* Olivia Langdon's commonplace book; quoted in *MC&MT*, 76.

78 *one of the greatest creations* SLC to Mollie Fairbanks, Elmira, Aug. 6, 1877, in *Mark Twain to Mrs. Fairbanks*, ed. Dixon Wecter (San Marino, Ca.: Huntington Library, 1949), 207. Twain said he had read Carlyle's book eight times ("My First Lie and How I Got Out of It," *CTSSE, 1891–1910*, 444). Paine's comment on Twain's deathbed reading of the book appeared in the *Hartford Courant* the day after Twain's death (Howard G. Baetzhold. *Mark Twain and John Bull* [Bloomington: Indiana University Press, 1970], 337). For more on Twain's responses to Carlyle throughout his life, see Wesley Britton, "Carlyle, Clemens, and Dickens: Mark Twain's Francophobia, the French Revolution, and Determinism," *Studies in American Fiction* 20, no. 2 (Autumn, 1992): 197–204; Howard G. Baetzhold, "Thomas Carlyle," *MTE*, 126–28; Alan Gribben, *Mark Twain's Library: A Reconstruction*, 2 vols. (Boston: G. K. Hall, 1980), 1:127–30; and Walter Blair, "The French Revolution and Huckleberry Finn." *Modern Philology* 55 (August 1957): 21–35.

78 *eight times* Samuel Langhorne Clemens to William Dean Howells, Elmira, Aug. 22, 1887, in *MTL* 2: 490.

79 *or to the portable hammock* Paine notes that Twain and his brother-in-law, Theodore Crane, "were always fond of each other and often read together the books in which they were mutually interested. They had portable-hammock arrangements [at Quarry Farm], which they placed side by side on the lawn, and read and discussed through summer afternoons." *Paine*, 1: 510.

79 *O poor mortals,* Thomas Carlyle, *The French Revolution*, vol. 1 (New York: Scribner, Welford, 1871), 160.

79 *Jervis Langdon had been a "conductor"* Ida Langdon wrote that Jervis Langdon, her grandfather, "was a 'conductor' in the Underground" (*Mark Twain in Elmira*, ed. Robert D. Jerome and Herbert A. Wisbey [Elmira, N.Y.: Mark Twain Society, 1977], 20). His involvement is corroborated by other sources as well.

80 *up from Chesapeake Bay* Winifred Eaton, quoted in Thomas E. Byrne, *Chemung County, 1890–1975* (Elmira, N.Y.: Chemung County Historical Society, 1976), 519.

80 *entirely penniless* "Underground Railroad Activities in Elmira," by Abner C. Wright. A biographical sketch of John W. Jones written in 1945 by Wright for the *Chemung Historical Journal*; rpt. Sept. 1968, Aug. 1985, and Nov. 1993. Subsequent citations refer to the latest reprinting.

80 *Jervis Langdon, who had made a fortune.* Anon, "Underground Railroad: Route to Freedom," from the *Elmira Sunday Telegram*, March 8, 1961; rpt. *Chemung Historical Journal*, 6, no. 4 (June 1961). In this article Thomas K. Beecher is also listed as a supporter. But it is Langdon's name that appears in William Siebert's list of Chemung County operators, along with Jones' (Siebert, *The Underground Railroad from Slavery to Freedom*). See also Wright, "Underground Railroad Activities in Elmira," 1757.

80 *Eight hundred fugitive slaves* Wright, "Underground Railroad Activities in Elmira," 1757. For more on Jones' underground railroad activities, see Gretchen Sullivan Sorin, "The Black Community in Elmira," in *A Heritage Uncovered: The Black Experience in Upstate New York, 1800–1915* (Elmira, N.Y.: Chemung County Historical Society, 1988), 11; and William Still, *The Underground Railroad* (Philadelphia: Porter and Coates, 1872), 43–44. Jones went on to serve as sexton of Woodlawn Cemetery, starting in 1864, where he was charged with burying the Confederate prisoners who died in the prison camp set up in Elmira. Before he was done, he had buried 2,973 Confederate soldiers, marking each grave with the appropriate information. "He kept an accurate record of the location of the grave and the name, company, regiment and state of every body he interred. Wooden headboards were replaced in later years with the stones now to be seen." (W. Charles Barber. "Elmira as Civil War Depot and Prison Camp." An address at the Civil War Forum, Steele Memorial Library, March 15, 1960; rpt. the *Chemung Historical Journal* [Aug. 1985]: 753–56, quote p. 756.) According to Helen Jerome, when members of the Daughters of the Confederacy made a trip to Elmira some years ago with an eye to removing the Confederate dead to a cemetery in the South, they changed their minds when they saw the care with which Jones had arranged graves and kept records. (Helen Jerome, taped interview by author, Elmira, N.Y. [Mark Twain Room, Gannett-Tripp Library, Elmira College], Sept. 11, 1995).

80 *Langdon, together with Hiram Crane* Susan Crane provided William Siebert with this information, which he did not end up including in his book on the underground railroad. Gretchen Sharlow, "Mark Twain's Elmira Home Circle of Abolitionists and Underground Railroad Conductors," unpublished paper.

80 *Prominent abolitionists* At the memorial service for Olivia Lewis Langdon, Jervis Langdon's wife, Rev. Thomas Beecher "reminded those gathered that the Langdons had always been 'at the service of fugitives from slavery,' and that the Langdon home had always been open to abolitionists such as Garrison, Phillips, Quincy, Johnson, Gerrit Smith, Foster, Frederick Douglass." Beecher, *Olivia Langdon Eulogy Memorial Pamphlet* 7 (cited in Skandera-Trombley, *Mark Twain in the Company of Women* [Philadelphia: University of Pennsylvania Press, 1994]). It is not known how often Douglass visited the Langdons. We know that he stayed with them when they lived in Millport, New York, in 1838, and an undated letter Douglass wrote to Charles Langdon in the early 1870s referred to his having dined thirty years earlier in his home (Douglass to Charles Jervis Langdon, quoted and commented on in *Bangor (M.) Daily Whig and Courier*, Aug. 23, 1872; cited in William McFeeley, *Frederick Douglass* (New York: W. W. Norton, 1991), 277.

80 *there is evidence that Langdon* "An intriguing notice in the *Elmira Advertiser* titled 'Noble Donation' and dated November 30, 1864, gives testimony to the Langdon family's support for abolition. The announcement calls attention to the 'noble donation' Jervis Langdon made to assist 'white refugees who are fleeing from the ravages of the Southern rebels.' Langdon, along with other Elmira personages such as the Reverend Thomas Beecher, Dr. Silas Gleason, and Judge Ariel S. Thurston, donated $650 to this cause. What makes the item noteworthy is that the collected funds were given to a certain Reverend E. Folsom, a hospital chaplain in Cairo, Illinois. This money very likely funded Underground Railroad activities in the area, and the 'white refugees' were possibly abolitionists who had been discovered and were forced to escape. At the time, Cairo was solidly pro-Confederacy, and activities such as those being carried out by Reverend Folsom would have been cloaked with great secrecy." Skandera-Trombley' *Mark Twain in the Company of Women*, 73.

80 *conducted a private school* "Underground Railroad: Route to Freedom," from the *Elmira Sunday Telegram*, March 8, 1961; rpt. in *Chemung Historical Journal* (June 1961): 862–63. See also Herbert A. Wisbey Jr., "Clarissa Thurston's Ladies Seminary," *Chemung Historical Journal* (June 1989): 3857–62.

81 *Miss Thurston's* Resa Willis, *Mark and Livy: The Love Story of Mark Twain and the Woman Who Almost Tamed Him* (New York: Atheneum, 1992), 18.

81 *After the war, Langdon MC&MT*, 77.

81 *That the using, holding, or trading* Eva Taylor. *A History of the Park Church* (Elmira, N.Y.: Park Church, 1961); quoted in Sorin, "The Black Community in Elmira," 9. On April 24, 1996, a program presented at Park Church on the occasion of its sesquicentennial included a talk by Elmira College senior Tyrone Marsh celebrating Park Church's abolitionist heritage and affirming the importance of that heritage as background for understanding Mark Twain's greatest novel. Personal communication, Tyrone Marsh, May 23, 1996.

81 *If I had never seen* Douglass quoted in *Letters-3*, 428. Douglass's letter was written from Rochester on Nov. 9, 1870; the Langdons' aid to Douglass is cited in a note.

81 *generous white friends* By the time of the Civil War about sixty black families lived in Elmira, "enough people to form a religious group.

Leaders like John Washington, Primus Cord, and Jefferson Brown worked with generous white friends to build a church at Dickinson and Fourth Sts., which they named for Frederick Douglass." Byrne, *Chemung County*, 519.

81 *Louis J. Budd suggests that* Louis J. Budd, *Mark Twain, Social Philosopher* (Bloomington: Indiana University Press, 1962), 87.

82 *either savage assaults* Paine, 1:400.

82 *Ah, well! Too bad* [Twain], editorial in *Buffalo Express*, Aug. 26, 1869, p. 2.

83 *slightly inebriated and noisy* Twain, "Life on the Isthmus." *Buffalo Express*, Oct. 4, 1870; rpt. Henry Duskis, ed. *The Forgotten Writings of Mark Twain* (New York: Philosophical Library, 1963), 310.

83 *engendered Anthony Walton's powerful book* Anthony Walton, "Prologue," *Mississippi: An American Journey* (New York: Alfred A. Knopf, 1996), xi-xii. The man in Twain's sketch had been "slightly inebriated and noisy"; the black teenager murdered by the police in the scene described by Walton's father "forty years ago" was slightly "high" and noisy as well. Twain wrote in 1870, "When 'Mr. Negro' did not move along quickly enough, the policeman "wasn't at all put out—he only persuaded him to go by jabbing the bayonet into the poor wretch's head with all his force, and then as the blood streamed over his face, striking him on the skull with the barrel of his musket." Twain said he believed the man was buried the next day. In Walton's father's recollection of the analogous incident from his own childhood, he and a high school friend, both around fifteen or sixteen at the time, "had been celebrating Christmas Eve, you know, blowing off some steam, and James was a little high, he was a little happy. We went into [the colored waiting room in] the bus station and James was singing and clapping his hands. He was just a boy, you know, having fun, but somebody called the sheriff and complained. I never found out who" (xi). The sheriff gratuitously "pulled his .38 and shot the boy in the back. He didn't say 'stop!'—nothing. . . . Right there. That's where he fell, dead, on Christmas eve" (xii). Twain's complaints about police brutality predate this sketch; see, for example, his satire on police brutality toward the Chinese in San Francisco entitled "What Have the Police Been Doing?" *San Francisco Golden Era*, Jan. 21, 1866; rpt. in *CTSSE, 1852-1890*, 196-98.

83 *In San Francisco, the other day*, Twain, "Disgraceful Persecution of a Boy," *Galaxy*, May 1870; rpt. in *CTSSE, 1852-1890*, 379-92, quote p. 379. Twain had explored this theme a few years earlier as well.

83 *It was in this way that the boy* Twain, "Disgraceful Persecution," 380.

83 *Defining right behavior and impeaching bad* Steinbrink, *Getting to Be Mark Twain*, 107.

84 *The 179 households* Byrne, *Chemung County*, 519.

84 *Mrs. Luckett was a slave* This comment in Clemens' handwriting appears on the flyleaf of his copy of William Still, *The Underground Railroad*, rev. ed. (Philadelphia, William Still 1883); cited in Gribben, *Mark Twain's Library*, 2: 666).

84 *Each story was different* The Chemung County Historical Society Black History Records contain numerous oral histories of underground railroad escapes collected from the ex-slaves' Elmira descendants. For example, Jenny Dunmeyer told an interviewer that her great grandfather arrived in the late 1830s. A tavern and livery-stable owner hid him in the hayloft in his barn for several days when his former master, who was also his father, arrived in Elmira with his overseer to drag him back to slavery. The barn proved a safe haven. He decided to stay in Elmira and accepted the stable owner's offer of a job. Jenny Dunmeyer, 314 Washington St., Elmira, N.Y., Oct. 13, 1989. Interview by Eva Williams. Chemung County Historical Society Black History Records, Oral Project, Box #1 (transcript, p. 1).

84 ELMIRAN, 80, FORMER SLAVE *Sunday Telegraph/Elmira Star-Gazette*, April 3, 1938, p. 1. Courtesy of Donald Blandford.

85 *kept right on going* Donald Blandford, taped interview by author, Elmira, N.Y. (Green Pastures), Sept. 12, 1995. See also ELMIRAN, 80, FORMER SLAVE.

85 *a house at 811 East Avenue that Mark Twain passed* Blandford interview. Donald Blandford recalls his grandfather having commented on Twain's passing by his house daily. The farm's natural beauty and its connection to the famous author attracted Alsace, a talented artist, who did oil paintings of Twain's hilltop study, of the Clemens family's engraved stone watering troughs, and of Twain's memorial to his daughter. Blandford's painting of one of the Clemens watering troughs is pictured in the 1938 *Elmira Star-Gazette* article noted earlier. Several of his paintings, his grandson said, are also owned by the Chemung County Historical Society and the Arnot Art Museum. Unfortunately, most of the hundreds of canvases he painted were destroyed, first by flood and then by fire. He earned his living as a painter and paperhanger.

85 *sitting by the Erie Railroad platform* William Condol's recollections of his grandfather, Henry Crummell Washington, are cited in Herbert A. Wisbey Jr., "The True Story of Auntie Cord." *Mark Twain Society Bulletin* (Elmira, N.Y.) 4, no. 2 (June 1981): 3–4.

85 *masons, teamsters, grocers* Thomas E. Byrne, "Elmira's Black History. *Chemung Historical Journal*, 14, no. 1 (Sept. 1968): 1749.

85 *Members of Elmira's elite* Wisbey, "The True Story of Auntie Cord." Wisbey notes that "when Henry Washington died, on Washington's birthday in 1927, at the age of eighty-two, he was the oldest barber in Elmira." In 1986 *Mark Twain Society Bulletin* editor Robert Jerome reprinted an 1886 article published by an anonymous reporter for *The Morning Telegram* (Elmira, N.Y.) on June 27, 1886, that described one of Twain's typical visits to the Rathbun Hotel barber shop (where Henry Washington worked as a barber at the time): "Mark Twain is heartily welcomed at the Rathbun house shop. . . . Mark's hair is a great source of trouble to the artist who has it in charge. By nature it resists all rules of the barber shop, and if that was not enough, no sooner is it plastered down in accordance with the line of beauty established by the tonsorial profession, then [*sic*] the owner, if he can get his hands from under the apron, runs his finger through it and it is again flowing to the wind." *Mark Twain Society Bulletin*, 9, no. 2 (July 1986): 1–2.

85 *she had the best gift of strong & simple speech* Twain, "A Family Sketch," p. 61, Mark Twain Collection, James S. Copley Library, La Jolla, Ca. Quoted with permission in *WHB*, 8.

85 *a shameful tale of wrong & hardship . . . a curiously strong piece of literary work* Twain, typescript of notebook 35, May–Oct. 1895, p. 8. Mark Twain Papers. Quoted with permission in *WHB*, 9.

87 *Dey put chains on us* Twain, "A True Story, Repeated Word for Word as I Heard it," in *Mark Twain's Sketches, New and Old*, OMT, 204.

87 *in slavery more than forty years* Twain, description of a photo of the family group at Quarry Farm that he inserted into a letter; quoted in *Paine*, 1:509.

87 *auction block came into my personal experience* A Connecticut Yankee, 346.

87 *She dropped on her knees* A Connecticut Yankee, 199.

88 *I knowed what it was about* Huckleberry Finn, 201.

88 *Negroes . . . are comparatively insensible to pain.* Anon., "Negro," *American Cyclopaedia: A Popular Dictionary of General Knowledge*, ed. George Ripley and Charles Anderson Dana, 16 vols. (New York: Appleton, 1873–76), 216. Clemens donated his copy of the encyclopedia to the Mark Twain Library in Redding, Conn., and consulted it often for information on American history and biography (Gribben, *Mark Twain's Library*, 1:23–25).

88 *Their griefs are transient* Thomas Jefferson, *The Writings of Thomas Jefferson*, definitive ed. Vol. 1, *Notes on Virginia*, ed. Albert Ellergy Bergh (Washington, D.C.: Thomas Jefferson Memorial Association, 1907), 194.

88 *the book's real message . . . The book [was] a cry from the heart* McFeeley, *Frederick Douglass*, 311–312.

89 *I was delighted when I found* Mary Ann Cord's granddaughter, Louise Florence Washington Condol (1868–1972), one of Henry Washington's seven children, who had played with Mark Twain's children as a child and who later became a leading figure in Elmira's black community, was Donald Blandford's great aunt (by marriage). Both she and Alsace Blandford married members of the Condol family, linking the Cords/Washingtons, the Blandfords, and the Condols in a close-knit extended-family network in Elmira.

89 *People would come in* Blandford interview.

89 *"We can't" was always the excuse* Robert Jerome, taped interview by author, Elmira, N.Y. (Mark Twain Room, Gannett-Tripp Library, Elmira College), Sept. 11, 1995.

90 *Blacks are tired, angry* Newsletter quoted from memory by Blandford during interview.

90 *You might be interested in these* Blandford interview.

91 *remembered playing with the Clemens children* Louise Condol, quoted by Thomas Byrne (from interview) in Byrne, *Chemung County*, 519. Her son, Leon W. Condol, interviewed her in 1964 for "Negroes Lived Here Long Before Civil War," an article that appeared in the centennial edition of the *Elmira Star-Gazette* (June 26, 1964). The article states that "Mrs. Louise Condol, who was 96 last May 26, is a living link with 19th-century Elmira. She was born in Elmira, the daughter of Mr. and Mrs. Henry Crummell Washington. Mrs.

Condol, who lives at 606 Dickinson St., used to sit on the steps of Mark Twain's study at Quarry Farm and listen to the author read to her and his children."

91 *"pump" scene in Tom Sawyer* Here Tom Sawyer "remembered that there was always company at the pump. White, mulatto and negro boys and girls were always there waiting their turns, resting, trading playthings, quarreling, fighting, skylarking." *Tom Sawyer*, 11.

91 *talking with Mark Twain on several occasions* James Lewis, quoted in Robert D. Jerome and Herbert A. Wisbey Jr., ed., *Mark Twain in Elmira* (Elmira, N.Y.: Mark Twain Society, 1977), 180. *Tom Sawyer* was published in Britain in June 1876 and in the United States in Dec. 1876.

92 *an almost mythical status* Henry Nash Smith, *Mark Twain: The Development of a Writer* [1962] (New York: Atheneum, 1974), 75.

92 *while some critics argue* For a historical overview of critical responses to the novel, see Gary Scharnhorst, "Introduction," *Critical Essays on the Adventures of Tom Sawyer*, ed. Gary Scharnhorst (New York: G. K. Hall, 1993), 1–17.

92 *pre-industrial pieties . . . visual rehearsal* Scharnhorst, "Introduction," 5.

92 *filmed in Technicolor* Scharnhorst, "Introduction," 5.

93 *"hymn" to boyhood* In *MC&MT* Justin Kaplan wrote, "*Tom Sawyer*, which Clemens once described as 'simply a hymn' to boyhood . . ."(197). Twain's refers to the book as a "hymn" in "Unmailed Answer," Sept. 8, 1887, *MTL* 2: 477.

93 *"Facts Concerning the Recent Carnival of Crime in Connecticut"* Mark Twain published this piece in *The Atlantic Monthly* 37 (June 1876): 641–50; rpt. in *CTSSE, 1852–1890*, 644–60.

93 *the dreaminess, the melancholy, the romance* Samuel Clemens to Will Bowen, Aug. 31, 1876, writing from Elmira. See *Mark Twain's Letters to Will Bowen* (Austin: University of Texas Press, 1941), 24.

93 *All this is* Samuel Clemens to Will Bowen, Aug. 31, 1876, writing from Elmira. See *Mark Twain's Letters to Will Bowen*, 24.

93 *As to the past* Samuel Clemens to Frank L. Burrough, Nov. 1, 1876. In this letter Twain describes comments he made previously to Will Bowen. See *Mark Twain's Letters to Will Bowen*, 9.

94 *Where is the use in bothering what a man's character was* Twain to the editor of the *New York World*, Feb. 14, 1877. The letter appeared in the *World* on February 18, 1877. The occasion of the letter was Clemens' belief that Captain Charles C. Duncan of the *Quaker City* had slandered him, and he "responded as Mark Twain, in a public letter to the editor of the New York *World*," from which this quotation comes. See Steinbrink, *Getting to Be Mark Twain*, 191, 210 (letter, rpt. p. 191).

94 *the most prominent and honored colored man in the world* Anon., "Emancipation. Why and How the Colored People Celebrate To-Day." *EA*, Aug. 3, 1880, p. 2.

95 *colored conventions* John Blassingame, ed. *The Frederick Douglass Papers*, vol. 2 (New Haven, Conn.: Yale University Press, 1982), 68. Cited in Sorin, "The Black Community in Elmira," 10.

95 *the anniversary of British liberation of slaves* Anon., "The Colored People: How they Observed the Emancipation Anniversary. A Large Crowd and a Big Success." *EA*, Aug. 4, 1880, p. 5. I am grateful to Victor Doyno for having made me aware of the fact that Douglass gave this speech in Elmira in August 1880.

95 *During the final week of the campaign* Paul Fatout, headnote to "Political Speech, Republican Rally, Hartford Opera House, Oct. 26, 1880," in *Mark Twain Speaking*, ed. Paul Fatout (Iowa City: University of Iowa Press, 1976), 138; the speech itself is reprinted on pp. 138–45. Fatout comments that Twain "talked at greater length than usual" and notes that the *Courant* "remarked that the audience was 'held unbroken to the very close, at 10 o'clock,' " (138). Twain, "Funeral Oration Over the Grave of the Democratic Party, Republican Jollificiation, Hartford Opera House, Nov. 2, 1880," in *Mark Twain Speaking*, pp. 146–48.

95 *But Twain did have an interest in Douglass* Mark Twain's first act of the New Year in January 1881 was to intercede in Washington to try to help Douglass keep his job as Marshal of the District of Columbia. "He is a personal friend of mine, but that is not the point," Clemens wrote Garfield, the newly elected President. Quoted in McFeeley, *Frederick Douglass*, 305.

95 *I would like to hear him make a speech.* Samuel Clemens to Olivia Langdon, Dec. 15–16, 1869, in *Letters-3*, 426.

95 *excitement reached the white folks* Anon., "The Colored People . . . ," 5. The procession included "Colored Veterans of the Late War," the "Reception Committee in carriages," "Distinguished Guests in carriages," "Thirty-eight young ladies representing the thirty-eight states," "Rescue Hook and Ladder Company of Norwich," "Horseheads Hose Company (colored)," "Elmira Colored Y.M.C.A.," and "Masonic and other civic societies." In "Emancipation: Why and How the Colored People Celebrate To-Day." *EA* Aug. 3, 1880, p. 2.

95 *a drum corps of small boys* Anon., "The Colored People . . . ," 5.

95 *"fairly alive" with spectators* Anon., "The Colored People . . . ," 5.

95 *sounds of the brass bands and drums* The parade route was as follows: "The procession will form on Dickinson street with the right resting on Clinton, at noon precisely. The line of march will be down Lake to Water; up Water to Main; up Main to Church; up Church to Walnut; up Walnut to Hoffman's Grove" ("Emancipation . . . ," 2). Olivia Lewis Langdon lived at the corner of Church and Main. It is the view of Michael Kiskis, a professor at Elmira College, who lives on Water Street, that the music of a brass band playing along the parade route in 1880 would easily have been heard at Quarry Farm (personal communication, Sept. 26, 1995). Notices about Douglass' visit to Elmira began appearing in the *EA* beginning on Saturday, July 31, when a paragraph headlined DOUGLASS COMING announced that "Frederick Douglass, the world-renowned colored orator . . . will reach Elmira Monday evening next" ("Douglass Coming," *EA*, July 31, 1880, p. 2).

96 *Committee on Arrangements and the Committee on Reception* "Emancipation . . . ," 2. John T. Lewis is among the thirteen men listed on the

"Committee on Arrangements" and is among the five men listed on the "Committee on Reception."

96 *As the venerated and noble colored man* "Emancipation . . . ," 2.

96 *Although particularly intended for* "The Address of Hon. Frederick Douglass." *EA*, Aug. 4, 1880, p. 1.

96 *To-day in all the Gulf States* Frederick Douglass, as quoted in "FRED DOUGLASS! His Great Speech Yesterday. His Political Views Plainly Expressed. A Splendid Campaign Document. Hancock's 'Statesmanship' Portrayed. Why ALL Voters should Support Garfield. READ AND CIRCULATE" (*EA*, Aug. 4, 1880, p. 2).

97 *The final portion of Huckleberry Finn* See *WHB*, 68–76. "Louis J. Budd suggested in 1959, and then again in 1962, that *Huckleberry Finn* may have been a comment not only on the 1830s and 1840s but also on the 1880s, and scholars have subsequently built an increasingly solid case for the idea that the last portion of the novel may be read as a commentary on American race relations in the post-Reconstruction era" (*WHB*, 74). See works on this theme by Tony Tanner, Victor Doyno, Neil Schmitz, Richard and Rita Gollin, Lawrence Holland, Stephen Mailloux, Harold Beaver, Charles Nilon and Peter Messent; (*WHB*, 219–247), and Messent, *Mark Twain*, forthcoming.

97 *As Twain struggled to complete* Victor Doyno has ably described the convict–lease system in the post-Reconstruction South (*Writing Huck Finn* [Philadelphia: University of Pennsylvania Press, 1991], 231–33) and has directed our attention to places Twain would have been likely to encounter discussions of it between 1883 and 1884. He writes, "Could Twain have learned of this injustice? He could have read a horrifying explanation about using chase dogs to track and a more fierce 'catch dog' to kill escaping prisoner-laborers in the January 1883 *Atlantic* 'Studies in the South' article entitled 'The Survival of Slavery.' But a much more knowledgeable and zealous personal informant existed. Twain met George Washington Cable in June of 1881 and subsequently visited him in New Orleans. The two men spent many days together during the period of 1882–1885, including a time when Cable was in Twain's home. [Cable] was already quite knowledgeable and indignant about the horrible evils of the convict–lease system when he spoke in New Orleans, as reported in the *Times Picayune*, on Jan. 9, 1883. He gave another lecture on the topic in New York on March 26, 1883, and later came to Hartford where he stayed with the Warners and visited Twain during April 2–5. . . . He read the proofs of his powerful essay, 'The Convict–Lease System in the Southern States,' on November 24, and this detailed, definitive essay appeared in the February 1884 *Century*. The same magazine would soon publish Cable's relevant 'The Freedman's Case in Equity.' The second article included two pages Cable had originally put in the 'Convict–Lease' article but suppressed" (233–34). These articles must have reinforced Twain's awareness of a problem that Douglass's speech in Elmira had alerted him to as early as 1880.

97 *It is one's human environment* Inscribed authorized edition of *The Works of Mark Twain* in Mark Twain Room, Gannett-Tripp Library, Elmira College.

98 *George Griffin, known sometimes as Twain's butler* The most extended profile

of him by Twain is in the unpublished "Family Sketch" in the Mark Twain Collection of the James S. Copley Library in La Jolla, California.

99 *He was renowned for being able to take the hats of four hundred men* Obituary for Warner McGuinn, the *Afro-American* (Baltimore), July 17, 1937 (extracts rpt. the *Baltimore Afro-American*, March 23, 1985, in an article entitled, "Is Huckleberry Finn Racist? Was Its Author Anti-Black? Warner T. McGuinn and Mark Twain.")

100 *the most grotesque example of racist trash ever written* Dr. John Wallace on ABC's *Nightline* broadcast entitled " 'Huckleberry Finn': Literature or Racist Trash?" *Nightline*, Ted Koppel anchor, Feb. 4, 1985, show #966 (transcript). John Wallace repeated the same charge in the same language ("the most grotesque example of racist trash ever written") ten years later. On this occasion he also asserted that "any teacher caught trying to use that piece of trash with our children should be fired on the spot, for he or she is either racist, insensitive, naive, incompetent or all of the above and has no business in a public-school classroom." John Wallace. "This Book Is Just Trash," *USA Today*, final edition, Aug. 25, 1995, p. 12A.

100 *[I] felt it was one of the best indictments against racism* Meshach Taylor on *Nightline*, Feb. 4, 1985.

101 *I think the book is certainly the most racist book* Dr. John Wallace on *Nightline*, Feb. 4, 1985.

101 *The result was a column* Shelley Fisher Fishkin, "Twain in '85," *NYT*, Feb. 18, 1985, p. A17.

101 *Dear Sir, Do you know him?* Samuel Clemens to Francis Wayland, Dec. 24, 1885. Boxed text of letter rpt. Edwin McDowell, "From Twain, a Letter on Debt to Blacks." *NYT*, March 14, 1985, p. C21. 1885 is the year of the book's publication in the U.S. It came out in England in 1884.

103 *The reply to Twain's letter* Francis Wayland to Samuel Clemens, Dec. 25, 1885. See Philip Butcher, "Mark Twain's Installment on the National Debt." *Southern Literary Journal*, 1, no. 2 (Spring 1969): 48–55.

103 *Most of the law school's nineteenth-century records* This account of my research on Warner McGuinn draws on a previously published essay: Shelley Fisher Fishkin, "Changing the Story," in *People of the Book: Thirty Scholars Reflect on Their Jewish Identity*, ed. Jeffrey Rubin-Dorsky and Shelley Fisher Fishkin (Madison: University of Wisconsin Press, 1996), 47–63.

103 *Born near Richmond, Virginia, in 1862* Various sources give different dates of birth for McGuinn. His obituary in the *NYT* says that he died in 1937 at age seventy-four, which would have made his birth date 1863. Philip Butcher says he was born in "1862 or 1863" ("Mark Twain's Installment," 50). *Biographies of Graduates of Yale Law School, 1824–1899*, ed. and comp. Roger W. Tuttle, lists his birth date as 1862. A typed biographical sheet supplied by Lincoln University that accompanied Senator Joseph I. France's letter about McGuinn to President Calvin Coolidge lists McGuinn's birth date as 1864.

104 *I think the colored youth is a promising case* Francis Wayland to Samuel Clemens, Dec. 25, 1885. Letter quoted by permission of the Mark Twain Project, Bancroft Library, Berkeley.

104 *very studious, & well behaved* Francis Wayland to Samuel Clemens, Dec. 30, 1885. Letter quoted by permission of the Mark Twain Project, Bancroft Library, Berkeley.

104 *the most promising colored youth we have ever had"* Francis Wayland to Samuel Clemens, Oct. 6, 1886. Quoted by permission of the Mark Twain Project, Bancroft Library, Berkeley.

104 *"The Constitutional Limitation to Land Taxes"* A list of four possible topics appears in "REGULATIONS (Law School Prizes and Theses.) Yale Law School, 1886-7. Subjects for Prize Orations, Essays and Theses." McGuinn's choice of this topic is specified in a typed biographical sheet in the possession of Lincoln University that accompanied the recommendation of Senator Joseph I. France of Maryland that President Calvin Coolidge appoint Warner McGuinn Recorder of Deeds for the District of Columbia (March 23, 1926).

104 *Your beneficiary of last year* Francis Wayland to Samuel Clemens, Oct. 1, 1887. Letter quoted by permission of the Mark Twain Project, Bancroft Library, Berkeley.

105 *One article described a near-riot in New Orleans* Anon., "Negroes On Stage. Stock Company Announcement Recalls Aldridge. Tragedian Passed Greater Part of Early Life Here and Was Booth's Valet." *Baltimore Sun*, July 16, 1916. Clipping in Warner McGuinn's scrapbook, Yale University Archives and Manuscripts.

105 *"constantly increasing" class* Eliot Norton, "For A Colored Army," Letter to the Editor, *Baltimore American*, Jan. 22, 1916. Clipping in Warner McGuinn's scrapbook, Yale University Archives and Manuscripts.

105 *Once the negroes began to join* "Negro Enlistment Feared." *Baltimore Morning Sun*, Jan. 15, 1916. Clipping in Warner McGuinn's scrapbook, Yale University Archives and Manuscripts.

105 *another clipping reported the cheers that greeted the introduction of two "Jim Crow car" bills Baltimore American*, June 16, 1916. Clipping in Warner McGuinn's scrapbook, Yale University Archives and Manuscripts.

105 *Arguing the celebrated Baltimore segregation case* Account of trial proceedings quoted in "Is Huckleberry Finn Racist? . . ."

106 *He was one of the greatest lawyers* McDowell, "From Twain, a Letter on Debt to Blacks."

107 *It's the smoking gun* Henry Louis Gates Jr. on *ABC Evening News*, Peter Jennings reporting, March 14, 1985.

107 *The best he could muster* John Wallace on "Freeman Reports," CNN, March 14, 1985.

108 *Dear Momma—Wherever you are Nigger: An Autobiography*, by Dick Gregory with Robert Lipsyte (New York: Pocket Books, 1964), [i].

108 *Last time I was down South* Gregory, *Nigger*, 144.

109 *all those Negro mothers.* Gregory, *Nigger*, 209.

110 *You folks know a lot* David Bradley, "The First 'Nigger' Novel." talk presented at the annual Meeting of the Mark Twain Memorial and New England American Studies Association, Hartford, Conn., May 1985.

110 *It made perfect sense* The only work of criticism up to this point that had

addressed the resonances between *Huckleberry Finn* and twentieth-century black writers in any depth was Arnold Rampersad's brilliant 1984 essay entitled *"Adventures of Huckleberry Finn* and Afro-American Literature," reprinted in *Satire or Evasion? Black Perspectives on "Huckleberry Finn,"* ed. James S. Leonard, Thomas A. Tenney, and Thadious M. Davis (Durham, N.C.: Duke University Press) 216–27.

111 *made it possible for many of us* Ellison interview by author.

111 *I think it comes full circle* Ellison interview by author.

111 *impudent and delightful and satirical young black man* Mark Twain, "Corn-Pone Opinions." (1901), in Twain, *Europe and Elsewhere* (New York: Harper and Brothers, 1923); rpt. in *CTSSE, 1891–1910*, 507–11; quote on 507. Jerry, the slave Twain recalls in the piece, had received virtually no critical attention. See *WHB*, 53–76, 173–74.

111 *the most artless, sociable, exhaustless talker* Twain, "Sociable Jimmy," *NYT*, Nov. 29, 1874, p. 7. See also *WHB*, 13–49; sketch rpt. on pp. 249–52.

111 *the greatest* man *in the United States* Mark Twain, manuscript of "Corn-Pone Opinions," The Mark Twain Papers, reproduced by permission on p. 56 of *WHB*.

112 *The spoken idiom of American negroes* Ralph Ellison, "What America Would Be Like Without Blacks," *Time*, April 6, 1970, p. 109, rpt. in Ellison's *Going to the Territory* (New York: Random House, 1987), 104–12.

112 *The black man* Ellison, "What America Would Be Like Without Blacks," 109.

112 *When we spoke on the phone* The conversation with Ralph Ellison took place in July 1992.

113 *sensed a black strain in Huck's voice* Hal Holbrook, quoted in Anthony De Palma, "A Black 'Voice' for Huck" (*NYT* wire story), *San Francisco Chronicle*, July 7, 1992, p. D2. (The Holbrook quote was not included in the version of this story that ran in the *NYT* but was in the one that ran in the *Chronicle*.)

113 *the genesis of Huck's* voice *and not his skin color* James R. Kincaid wrote that "if one registers Huck as black in any literal way, much of the powerful moral and social irony of the story disappears and some central scenes become nonsense." "Voices on the Mississippi," *NYT Book Review*, May 23, 1993, p. 12.

113 *How can Huck's voice be black if a sizable portion of it comes from white humorists?* This line of argument was pursued by Hamlin Hill in a review in *American Literary Realism* 26, no. 3 (Spring 1994): 90–92.

113 "What Every American Needs to Know" *Dictionary of Cultural Literacy*, ed. E. D. Hirsch Jr., Joseph Kett, and James Trefil (Boston: Houghton Mifflin, 1988).

114 *each approved textbook* amendment quoted in Tom Kennedy, "The Political Correctness of 'Huck Finn,' et al." *Houston Post*, April 5, 1995, p. A17.

115 *the mandatory reading* Senator Royce West, taped interview by the author, Dallas, Texas (Senator's law offices), Oct. 10, 1995. All of Senator West's comments quoted here were made during this interview.

115 *By the twisted standards* Robert T. Fagan, "The Academic Adventures of 'Huck Finn' " (Letters to the Editor), *WP*, March 20, 1995, p. A16.

116 *historical materials* Littlefield, quoted in J. Evetts Haley, *George W. Littlefield, Texan* (Norman: University of Oklahoma Press, 1943), 259–61.

117 *who fell in defence of law and order against fanaticism* Following the Equator, OMT, p. 321.

117 *Twain both began and abandoned this book* . . . For an extended discussion of this line of argument, see *WHB*, 69–75.

118 A COURT HOUSE FIRED . . . Headlines quoted by Twain in letter addressed "To the Editor of 'The Daily Graphic,' in New York City," Apr. 17, 1873, in *MTL*, 1:205. Twain's decision to clip the KLU KLUX MURDERS headline underlines his awareness of Klan activities during this period. Tom Perry suggests that we might "try reading the new cadaver scene" in the recently published first half of the manuscript of *Huckleberry Finn* as "a grotesque parody of the Ku Klux Klan," adding that this interpretation "adds force to the argument that the last chapters of the book present an allegory of the Reconstruction." Perry points out that "in the story Jim tells a white-hooded (the sheet around his head) corpse of a white man straddles and pins him (a black man) to the floor, leading to Jim's perfect punch-line and moral assessment, 'It warn't no way for a dead man to act, nohow; it might a scairt some people to death.' It's a fair description of the Klan's intent and modus operandi. This is all couched within a larger argument between Huck and Jim about whether the tale is even a ghost story or not. Huck's next line is, "But Jim, he warn't rightly a ghost—he was ony a dead man.' The key to the argument is the Klan's contention that they were the 'ghosts of the Confederacy,' their hoods worn to frighten superstitious blacks as well as to conceal their identities from the occupying federal troops. So Huck's line is a similarly accurate assessment and dismissal of the Klan acting in the name of the dead Confederacy. It's also a passing reference to the Klan itself being (well, almost entirely) dead, at least in its first incarnation, in 1876 when Twain was writing." Perry suggests that developments in the 1880s make "Twain's concern for the institutionalized repression of freed blacks, the subject he seemingly addresses at the end of the book, more pertinent and plaintive than the flurry of violent intimidations practiced by the Klan and parodied in the Cadaver scene. This reading . . . seems to indicate that Twain was interested in tackling the subject of the Reconstruction fairly early in the writing of the book and he may have removed this scene because the last chapters of the novel make it redundant." I am indebted to Perry, assistant director of publicity for Random House, for sharing this striking and innovative interpretation with me (letter to the author, May 23, 1996). See pp. 62–65 of the 1996 Random House edition of Twain's *Huckleberry Finn*.

118 *slavery to individuals* Frederick Douglass,"The Work of the Future," *Douglass's Monthly*, November 1862; rpt. Philip S. Foner, ed., *The Life and Writings of Frederick Douglass*, vol. 3 (New York: International Publishers, 1952), 292.

119 *The slave went free* W. E. B. Du Bois, *Black Reconstruction in America: An Essay Toward a History of the Part Which Black Folk Played in the Attempt to Reconstruct Democracy in America, 1860–1880* (New York: Harcourt, Brace, 1935), 30.

121 *articles in the press* See, for example, Valerie Strauss, "Twain Classic

Bounced From Class Again," *WP*, March 4, 1995, p. 1A; and Jonathan Yardley, "Huck Finn and the Ebb and Flow of Controversy," *WP*, March 13, 1995, p. D1.

121 *It's like taking a big part* Bill Matory, quoted in Nat Hentoff, "The Trials of 'Huckleberry Finn' " ("Op-Ed" Page), *WP*, March 18, 1995, p. A7.

121 *It would be a travesty* Jocelyn Chadwick Joshua recalled her remarks at the 1993 public meeting during an interview by the author, Nov. 12, 1995, Pittsburgh, Pa.

121 *In the high school* Jocelyn Chadwick Joshua, "Jim: Nobody's Fool," paper presented at the Dallas Public Library, Dallas, Texas, October 9, 1996.

122 *Irreverence, Twain once wrote MTNJ-3*, 392.

123 *If we'd eradicated the problem of racism* David Bradley, at Mark Twain Teachers' Institute, Hartford, Conn., July 24, 1995.

124 *saw nothing but all manner of pretty pictures* Twain, *Life on the Mississippi*, OMT, 118–121.

RIPPLES AND REVERBERATIONS

128 *Known To Everyone and Liked By All* See photo of "Mark Twain Cigar" wrapper in John Seelye, *Mark Twain in the Movies: A Meditation with Pictures* (New York: Viking, 1977), 10.

128 *He made personal appearances in* "Silver Legacy Announces Grand Opening, PR Newswire, July 13, 1995; Mel Shields, "Singer Christopher Is Sailing Toward a Comeback." *Sacramento Bee*, July 23, 1995, p. EN21; Chris Burbach, " 'The Most American Thing' In Tent Shows," *Omaha World Herald*, July 9, 1995, p. 1B; "Soirees Bring Power Singles Together," *Hartford Courant*, July 24, 1995, Business Weekly, p. 22; "Weekend Events," *Charleston Gazette* July 22, 1995, p. 3D; Robert Trussell, "If It Played Here, He Probably Had Something to Do With It," *Kansas City Star* July 16, 1995, p. J5; Mary L. Sherk, "Outdoor Pageants Bring History to Dramatic Life on a Grand Scale," *Denver Post*, July 16, 1995, p. T10; John Karras, "Muscatine's Red Carpet Has Only a Few Wrinkles," *Des Moines Register* July 30, 1995, Metro, p. 4.

128 *He contributed to* Robert J. Samuelson, "Requiem for the Typewriter," *Newsweek*, July 17, 1995, p. 43; Debra Gordon, "The Information Highway: What's Cooking in Cyberspace," *Virginian-Pilot* (Norfolk), July 16, 1995, p. F1; Stephen Harriman, "Sink Your Sweet Tooth into Lubeck, City of Marzipan and Medieval Spires," *Virginian-Pilot* (Norfolk), July 30, 1995, p. E1; Peter Berlin, "Safe in Hawaii's Comfort Zone," *Financial Times* (London), July 29, 1995, p. 1X; Tony Snow, "Next Year's Budget Still Bloated," *Cincinnati Enquirer*, July 30, 1995, p. E2; Randy Mink, "If the Heat Drives You Batty, Go Below," *Dallas Morning News*, July 23, 1995, p. 1G; Cynthia V. Campbell, "Rolling on the River; the Largest Sternwheeler in the World to Cruise America in Opulence," *Sunday Advocate* (Baton Rouge), July 30, 1995, p. 12; Critic-at-Large, "Dictionary Zenith on English Language," *San Diego Union*, July 23, 1995, p. E3; Steve Jarrett, "In Hollywood, Much Ado About Camelot," *News & Record* (Greensboro, N.C.) July 28, 1995, p. R91; Peter Truell, "Some Big

Funds, Like Soros's, Have Difficulty Despite Trend," *NYT*, July 27, 1995, p. D2; Franxis X. Clines, "An Ode to the Typewriter," *NYT*, July 10, 1995, p. D4; Margaret Vaughn, (no headline), *The Herald* (Glasgow) July 15, 1995, p. 17; "The Crisis and the U.S. Mission in Bosnia," CNN News (Domestic), late edition, July 16, 1995. (transcript #93); Andrew Cohen, "Lloyd Scanlan Suffers Setbacks," *Wisconsin State Journal*, July 29, 1995, p. 5d; "The Roots of Rush," *People*, July 24, 1995, Scene, p. 166; "Father Faints at Return of Egyptian 'Tom Sawyer'," Reuters, BC Cycle, July 24, 1995; "Foreign Press Center Briefing with Winston Lord," Federal News Service, State Department Briefing, July 24, 1995; Nancy Pate, "New Books Stretch Boundaries of Classic Arthur Tale," *Orlando Sentinel*, July 23, 1995, p. F1; "Judy Collins Newest Celebrity to Pen Novel" on "Showbiz Today," CNN, July 10, 1995 (transcript #843-44); "Unions Attack Safety Study," *Engineering News-Record*, July 14, 1995, 235, no. 4, p. 9; anon., "Fee Makes an Impact on Readers," *Orlando Sentinel*, July 23, 1995, p. G2; "Prepared Testimony of Dennis Karjala Before the House Judiciary Committee Subcommittee on Courts and Intellectual Property Hearings on H.R. 989," Federal News Service, July 13, 1995; Eric Adler, "Some Voices of Freedom," *Kansas City Star* July 4, 1995, p. E1; "Party Politics; NYPD Blues; One Nation After All," *MacNeil/Lehrer NewsHour*, July 4, 1995 (transcript #5262); Vern Anderson (AP), "Inmates Challenged to Explore Books, Set Minds Free . . . ," *LAT*, July 9, 1995, Metro, pt. B, p. 1; Evan Berland (AP), " 'Huckleberry Finn' Defenders Come to Classic's Rescue," *Wisconsin State Journal*, July 21, 1995, p. 8A.

129 *Twain's image has been used to sell* Copies of these advertisements are in the "Advertisements" file of the Mark Twain House.

129 *Though Mark Twain was fond of fowl* Advertisement quoted and described in Louis J. Budd, "Mark Twain as an American Icon," *CCMT*, 1-2. This splendid essay—along with Budd's magisterial *Our Mark Twain*—is a crucial starting point for anyone interested in Mark Twain's place in American popular culture. Also of interest is Louis J. Budd, "A 'Talent for Posturing': The Achievement of Mark Twain's Public Personality" in *The Mythologizing of Mark Twain*, ed. Sara deSaussure Davis and Philip D. Beidler (University, Alabama: University of Alabama Press, 1984), 77-98.

129 *Many a small thing*, A Connecticut Yankee, 210-11.

129 *Plumbers, Steam and Gas Fitters* cited in Budd, *Our Mark Twain*, 61.

129 *I think it is time the name* Twain, quoted in Budd, *Our Mark Twain*, 77.

131 *More fluid than a mosaic* Budd, "Mark Twain as an American Icon," 11-12.

131 *The Twain icon* Budd, "Mark Twain as an American Icon," 5.

131 *postured superbly, holding the spotlight* Budd, *Our Mark Twain*, xiii.

131 *pompano* Twain's paean to the delectable pompano fish he ate in New Orleans appears in *Life on the Mississippi*, OMT, 445.

133 *Mark Twain showed a sovereign contempt* George Gurley, "The Joy of Learning Unfolds Through Literature" *Kansas City Star* Sept. 21, 1995, p. F1.

133 *Mark Twain Community College* See Katherine Farrish, "College Won't Be

Named For Mark Twain." *Hartford Courant*, Dec. 22, 1992, p. D9; and Nat Hentoff, "Mark Twain and the Racists," *Sacramento Bee*, Jan. 5, 1993, p. B6.

133 *Mark Twain is cited* Kristen McNutt, "Let's Lighten Up; Leavening Nutritional Advice with Humor." *Nutrition Today*, 29, no. 3 (May 1994): 36; Beth Piskora, "Travelers-Check Players Taking Different Roads." *American Banker*, Apr. 13, 1995, p. 14.

133 *He is claimed as a forbear by* R. Z. Sheppard, "High Diddle-Didling: *The Confidence Man in American Literature,* by Gary Lindberg," *Time*, Dec. 28, 1981, p. 68; Alice Noble, "The Joy of Pessimism: You're Usually Right," UPI, Apr. 8, 1983; Rochelle O'Gorman Flynn, "The Spoken Word: Political Humorists Help Us Laugh Through Pain of Taxes." *Boston Herald*, Apr. 14, 1996, p. 52; Richard Benedetto, "Today's Political Jokes Show Respect Is a Lost Art," *USA Today* Apr. 1, 1996, p. 9A; Richard Rothschild, "Political Humor Used to Be Witty and Urbane . . . ," *Chicago Tribune*, Mar. 16, 1996, Tempo, p. 1; Paul Kurtz, "Two Scholars Debate: Who Needs God? Common Moral Decencies Don't Depend on Faith," *Dallas Morning News*, Oct. 28, 1995, Religion, p. 1G; Sandra Dickson, "Pets and People," *Dallas Morning News*, Oct. 21, 1995, p. 10C; John Blades, "An Author's Guide: Evanstan Publishing's Dorothy Kavka Shares the In's of Self-Publishing," *Chicago Tribune*, Jan. 12, 1995, Tempo, p.10B; Mary Wade Burnside, "Healthy Balance Homeopathy, West Virginia-Style." *Charleston Gazette*, Apr. 10, 1994, p.1D.

133 *He is summoned as an expert on* Anna Pavord. "Brassicas to Crow About . . . ," *The Independent*, Mar. 25, 1995, Gardening, p. 30; "75 Books Every Writer Must Read," *Writer's Digest*, 75, no. 12 (Dec. 1995): 38; Andrew Cohen, "Lloyd Scanlan Suffer Setbacks." *Wisconsin State Journal*, July 29, 1995, p. 5D; Lawrence S. Speidell, "Embarrassment and Riches: The Discomfort of Alternative Investment Strategies," *Journal of Portfolio Management*, (Fall 1990): 6.

133 *An American loves his family.* Twain, "Dinner Speech. Lotos Club Dinner for Mark Twain, January 11, 1908," in *Mark Twain Speaking*, 606.

133 *If caulifower, as Mark Twain wrote* Anna Pavord, "Brassicas to Crow About," p. 30. The Twain quote comes from *Pudd'nhead Wilson*, OMT, 67.

133 *Put all your eggs in one basket* Paul A. Samuelson, "The Long-Term Case for Equities." *Journal of Portfolio Management*, (fall 1994): 15. "Creating a Great Workplace," *Inc.* (Aug. 1994), 11; Sherwood Bass, "Employee Ownership Boosts Missouri Bank's Growth," Reuter Business Report, July 11, 1994; Tom Swain, "Economists Quantify Risk by Return Rate," *State-Times/Morning Advocate* (Baton Rouge), June 4, 1994, Business, p. 2C. The Twain quote comes from *Pudd'nhead Wilson*, OMT, 161.

133 *The foreign policy community* Federal News Service, "Hearing of the Commerce, Justice, State and Judiciary Committee of the House Appropriations Committee," May 8, 1996; Federal News Service State Department Briefing. "Foreign Press Center Briefing with Winston Lord," July 24, 1995; U.S. Department of State Dispatch, "Building a Pacific Community," Jan. 16, 1995, vol. 6, no. 3; Federal News Service State Department Briefing "Asia Society

Luncheon," Oct. 28, 1994. Mark Twain quotes Bill Nye in *Autobiography* (New York: Harper & Brothers, 1924), 1:338.

133 *Congress-bashers* Martin D. Tullai, "The Sorry State of Congresss," *Baltimore Sun*, Oct. 21, 1994, p. 15A; Jim Massie, "Book Offer Rings Up No Sale," *Columbus Dispatch*, Mar. 23, 1996, p. 2H; "Thoughts on the Business of Life," *Forbes*, June 19, 1995, p. 274; "Subbing Witlessness for Wit," *Wisconsin State Journal*, Feb. 10, 1995, p. 11A; "Congressional Records," *Orlando Sentinel*, June 11, 1994, p. A16. The Twain quotes come from *What Is Man?*, OMT, 106, and (Twain quoted in) *Paine*, 2:724.

133 *Get your facts first* John H. Senseman, "TCT and Stone Distort Nixon," *Capital Times* (Madison, Wis.) Jan. 13, 1996, p. 6A; Scott Montgomery, "Signing Off: In Search of a Cyber Identity," *Palm Beach Post*, Mar. 12, 1995, p. 1D; Robert S. Burk, "The Work of the ICC 'Dinosaur' " (Letters to the Editor), *WP*, July 30, 1994, p. A16; Mark Twain was quoted as having said this by Rudyard Kipling, who published the comment in an Indian periodical in 1890. For Kipling's account of the occasion on which Mark Twain made this remark, see *Mark Twain in Elmira*, ed. Robert D. Jerome and Herbert A. Wisbey Jr. (Elmira, N.Y.: Mark Twain Society, 1977), 102–110; quote p. 110.

134 *James Ross Clemens, a cousin of mine* White's recollections of Twain's remarks appear in "Mark Twain as a Newspaper Reporter," *Outlook*, 96 (Dec. 14, 1910): 964–65.

134 *Of course I'm dying* White, "Mark Twain as a Newspaper Reporter," 964–65.

134 *I sent a despatch* White, "Mark Twain as a Newspaper Reporter," 964–65.

134 *To paraphrase Mark Twain, stories about the death of newspapers. . . .* Roger Fidler, quoted in Patrick Greenlaw, "GOES–1 Weather Satellite Launched Successfully" on "Science & Technology Week," CNN, Apr. 16, 1994, (transcript #214).

134 *Despite the disappearance of some of the old familiar favorites* Caroline Perkins, "Seafood Sales 101: Get a Hook on the Basics," *ID: The Voice of Foodservice Distribution*, Sept. 1, 1994, vol. 30, no. 10, p. 54.

134 *Reports of the death of* Jerry Magee, "Mall's a Wild Card Joker, but La Mesa Player Didn't Fold," *San Diego Union-Tribune* Aug. 31, 1994, p. D7; "AT&T Makes the Smallest, Fastest Chip." *MASS HIGH TECH*, Feb. 7, 1994, vol. 12, no. 4, sec. 1, p. 14; "Access to Information: Effective Client/Server Mainframe Solutions," *Datamation*, June 1, 1993. vol. 39, no. 11, p. S3; Andy Pargh, "Satellite Dish Accessories," *St. Petersburg Times*, Sept. 24, 1989, p. 6H; "Videoconference 'Co-op' Looks Like Key to Success," *Data Communications*, April 1985, Newsfront, p. 60; Jeff Greenfield, "Payoffs Still Define Politics in Some Cities, " *LAT*, Apr. 13, 1986, Opinion, p. 3; Eloise M. Starbuck, "Year-End Surge Fortifies Local Residential Market," *Birmingham Business Journal*, vol. 13, no. 1, sec. 1, p. 18; "Shorthand," *NYT*, Apr. 6, 1986, sec. 3, p. 22. Babette Morgan, "Don't Slack Off, CPA Warns . . ." *St. Louis Post-Dispatch*, Jan. 15, 1993, Business, p. 1B; Federal News Service Defense Department Briefing, "Defense Secretary William Perry Addresses a Combined Meeting of the Mid-America Committee, the Chicago Council on Foreign Affairs, and the Chicagoland Chamber of Commerce," Dec. 11, 1995; Robert Luke, "Personal

Business Insider Trading: As Utilities' Stock Prices Fall, Buying Activity Is Picking Up," *Atlanta Journal and Constitution*, Nov. 28, 1994, Business, p. E8; Wayne Thompson, "Cassette Deck Is Still the One," *Oregonian*, Mar. 15, 1995, p. E2; Kevin Cullen, "Suspected IRA-Killing Is Said Unlikely to Derail Peace Process," *Boston Globe*, Nov. 13, 1994, p. 19; Dan Fisher, "Traditions Under Attack; Big Changes Brewing for British Pubs," *LAT*, May 27, 1989, p. 1; Steve Kay, quoted in Mike Dennis, "Battle for Britain: Cola Soft Drinks," *Super Marketing*, Mar. 31, 1995, no. 1166, p. S20; Eddie George, quoted in Richard Murphy, "Economist Proclaims Death of Inflation, Bank Unimpressed," *Reuter European Business Report*, Apr. 17, 1996; Stephen J. Ackerman, "Measles on the Rebound," *FDA Consumer* 20 (Oct. 1996):18; Joe Donnelly, "Surviving to Tell Another Tale," *The Herald* (Glasgow) Aug. 12, 1995, p. 12; Larry Korb, "U.S. Arms Industry Keeps on Rollin,' " *The Christian Science Monitor*, Jan. 30, 1992, p. 18; Beth Piskora, "Travelers-Check Players Taking Different Roads," *American Banker*, Apr. 13, 1995, 14; Jordan B. Moss, "Yiddish Very Much Alive And Kicking," *Newsday*, Feb. 3, 1994; NBC Olympics coverage, July 27, 1996, 3:40 EST.

135 *it was not that Adam ate the apple for the apple's sake* Twain, *Mark Twain's Notebook* (1935 ed.), 275; quoted in *MTAYF*, 5.

135 *patriotism is usually the refuge of the scoundrel* Twain, "Education and Citizenship," *Mark Twain's Speeches* (1923 ed.), 378; quoted in *MTAYF*, 350.

135 *Few things are harder to put up with* Twain, "Pudd'nhead Wilson's Calendar," *Pudd'nhead Wilson*, chap. OMT, p. 246.

135 *To get the right word in the right place* Samuel L. Clemens to Emeline B. Beach, Feb. 10, 1868 (Washington, D.C.), in *Letters-2*, 182–83.

135 *To condense the diffused light of a page of thought* Samuel L. Clemens to Emeline B. Beach, Feb. 10, 1868 (Washington, D.C.), in *Letters-2*, 183.

135 *A powerful agent is the right word* Mark Twain, "William Dean Howells," in *What Is Man? and Other Essays* (1917 ed.), 229, quoted in *MTAYF*, 521.

136 *What a bare, glittering ice-berg is mere intellectual greatness* Twain's marginal comments in Thackeray's "Swift" (1935 ed.), p. 55, quoted in *MTAYF*, 207.

136 *The answer to all of these questions* Roland De Wolk, "Twainisms That Ain't: Hannibal Sage Gets Credit for Too Much." *Oakland Tribune*, Aug. 4, 1991, pp. A3–4.

136 *Here are some things he didn't say that the editors* Roland De Wolk, "Twainisms That Ain't," pp. A3–4. See also Ken Kashiwahara, interview with Robert Hirst on "World News Tonight with Peter Jennings," ABC News, Oct. 22, 1991; and Ralph Keyes, "The Twain Syndrome," in *"Nice Guys Finish Seventh": False Phrases, Spurious Sayings, and Familiar Misquotations* (New York: HarperCollins, 1992).

137 *Two of the most famous quips widely attributed to Twain* Atribution to Warner in Seelye, *Mark Twain in the Movies*, 49.

137 *Huckleberry Finn is dead* Jim Goldberg, quoted in "Runaway's World Not Like Huck Finn," *Rocky Mountain News*, Oct. 1, 1995, p. 5A.

137 *It's no longer like Huckleberry Finn* Nancy Matthews, quoted in Sandra

Mathers, "New Center Designed to Help Runaway, Homeless Youths," *Orlando Sentinel*, Oct. 1, 1995, p. K1.

138 *Rosie the Ribiter and Heavy Metal* "Jumping Frog Breaks World Record," May 18, 1986; "Heavy Metal Wins Frog Jump," May 22, 1989. Attendance at the 1996 event was 37,000. Jonathan Rabinowitz, "Will Twain Stop Traffic in a 'Pass-Through City'?", *NYT*, Aug. 18, 1996, p. A17.

138 *deformed and mutilated* Clara Clemens Samoussoud's lawsuit against Columbia Pictures is quoted from and described in the following undated and unidentified newspaper clippings in the "Notes and Clippings on Mark Twain" file of the James S. Copley Library, La Jolla, Ca.: "Suit Charges Film 'Mutilates' Twain's Jumping Frog Story"; "Mark Twain's Daughter Sues on 'Corny' Film"; "Studio Sued Over 'Mutilation' of Twain Frog Tale"; and "Suit Charges Film 'Mutilates' Twain's Jumping Frog Story." The articles, which refer to the film *Best Man Wins* as having been released the previous year, are apparently from 1949. The seventy-eight-minute black-and-white film, directed by John Sturges and written by screenwriter Edward Heubach, was released in 1948.

139 *Huckleberry Finn, If I were Huckleberry Finn* "Huckleberry Finn," by Cliff Hess, Sam. M. Lewis, and Joe Young, (New York: Waterson, Berlin & Snyder, 1917) (sheet music in possession of the author). The song was recorded by Sam Ash (Columbia), Prince's Band (Columbia), the Little Wonder Band, and Gus Van and Joe Schenck (Victor); it was revived by Guy Lombardo and His Royal Canadians (see Barbara Cohen-Stratyner, *Popular Music, 1900–1919* [Detroit: Gale Research, 1988]132–33.) For information on Lewis, Hess, and Young, see entries in *The ASCAP Biographical Dictionary of Composers, Authors, and Publishers* (New York: ASCAP, 1966), 331, 440, 809–10; David A. Jason, *Recorded Ragtime* (Hamden, Conn.: Archon Books, 1973), 30, 86; and Gerald Boardman, *American Musical Theatre: A Chronicle* (New York: Oxford University Press, 1992), 313, 331.

139 *The same popular image* James A. Miller, professor of American studies at Trinity College, Hartford, recalled his surprise at finding himself referred to in this manner after being admitted to Brown University. He made this comment in a talk he presented at the Teachers' Institute, held in Hartford's Aetna Center under the auspices of the Mark Twain House, Hartford, Conn. July 27, 1995.

139 *A more nuanced version* For more on Weill's work on this project, see Jurgen Schebera, *Kurt Weill: An Illustrated Life* (New Haven, Conn.: Yale University Press, 1995). Weill's songs were performed in a memorial tribute to him held in New York in 1953. In August 1994 they were performed at a "Great American Concert" at the Hollywood Bowl under the direction of John Mauceri. Friday and Saturday night performances were attended by over thirty-one thousand people. (See Don Heckman, "Pop Music Review," *LAT*, Aug. 15, 1994, p. F5; and Shauna Snow, "Morning Report," *LAT*, Aug. 11, 1994, p. F2.) Most recently, Weill's songs were performed in June 1996 in a cabaret show at the piano bar and cabaret Don't Tell Mama in New York City. (reviewed by Chip Deffaa, "A Lighter Side of Weill," *New York Post*, May 25, 1996).

139 *Operas, oratorios, orchestral suites, and choral works* An engaging one-act

opera based on "The Celebrated Jumping Frog of Calaveras County" (music by Lukas Foss, with a libretto by Jean Karsavina) toured the New England region in the fall of 1994, playing to an estimated thirty thousand schoolchildren (the opera was originally developed some years earlier by Sarah Caldwell). "Mark Me Twain," an opera by composer Bern Herbolsheimer and librettist Ben Shallat, set in Virginia City, had its world premiere in April 1993 with the Nevada Opera. A "War Prayer Oratorio" (music by Herbert Haufrecht, libretto by Mark Twain) had its world premiere on April 22, 1995, at the Broadway Theater in New York City, with the Ulster Choral Society and the Bach–Handel Festival Orchestra conducted by Lee H. Pritchard. "Mark Twain: Portrait for Orchestra," composed by Jerome Kern, was recorded on Columbia Records in 1956. Paul Alan Levi's "Mark Twain Suite," which was given its premiere by the New York Choral Society in Carnegie Hall in 1983, included Huck Finn's "Sunrise on the Mississippi" (featuring "bird calls and water ripples evoked by a banjo, an ocarina and a slide whistle, delicately coloring unison choral descriptions"), "The Great Joust" from *A Connecticut Yankee* ("set with a medieval cantus firmus and jazzy drama"), and "The Awful German Language" ("marked 'Andante con Mahler,' and [invoking] all the weighty Germanic complexity that Twain found in the language"). "The work concluded with a bluesy, playful retelling of Twain's ghost story, 'The Golden Arm.' " Edward Rothstein, "Music: New 'Twain Suite,' by Levi, Has Its Premiere," *NYT*, May 4, 1983, p. C18. See also Ellen Pfeifer, "Twain's 'Frog' Leaps into Boston as Opera," *Boston Herald*, Nov. 24, 1994, p. 57; Jack Neal, "Herbolsheimer: Mark Me Twain (Pioneer Center, Reno, Nevada)" *Opera News*, 58, no. 4 (Oct. 1993): 44.

139 *Singer-songwriter Jimmy Buffett's* In a note he sent to Gretchen Sharlow, director of the Center for Mark Twain Studies in Elmira, in August 1995 (which Sharlow graciously shared with me), Buffett acknowledged Twain as an important influence, particularly in his 1995 CD *Barometer Soup*.

139 *Stage productions* For an overview of stage and screen productions of Twain's works, see Wesley Britton, "Media Interpretations of Mark Twain's Life and Works," in *MTE*, 500–504. R. Kent Rasmussen provides information about media interpretations of Twain's works under the listings for many specific works in *Mark Twain A to Z: The Essential Reference to His Life and Writings* (New York: Facts on File, 1995; Oxford: Oxford University Press, 1996). The most extensive treatment of film versions of any single work is Clyde V. Haupt, *Huckleberry Finn on Film: Film and Television Adaptations of Mark Twain's Novel, 1920–1993* (Jefferson, N.C., and London: McFarland, 1994).

140 *split-screen technique* Rasmussen, *Mark Twain A to Z*, 370.

141 *It's all very good-natured silliness* Kurt Nicewonger (United Feature Syndicate), " 'Prince' Could Make Twain Roll Over in His Grave," *Fresno Bee*, Oct. 6, 1995, p. F11.

143 *"Wishbone," a Jack Russell terrier* Anon., "Every Dog Has His Daydream: Inteview with Wishbone the Jack Russell Terrier," *Daily Mirror* (London), Dec. 30, 1995, Features, p. 3.

143 *twenty-five thousand "Tom" hopefuls* A photograph from David O. Selznick's publicity department (accompanied by a letter dated Oct. 22, 1941) showing the fence-painting scene from the film hangs in the glass-covered display boards at the Mark Twain Boyhood Home and Museum in Hannibal, with the following caption (apparently written by the studio and sent with the photograph): "The quaint little Mississippi village of St. Petersburg, which was really Mark Twain's home in Hannibal, has been reconstructed for the making of the Technicolor production of The Adventures of Tom Sawyer, in which David O. Selznick will introduce his latest discovery, 12-year-old Tommy Kelly, who has the starring role. An unknown lad without previous screen experience, Tommy was found in a nationwide talent search in which more than 25,000 boys were interviewed, viewed or tested."

144 *As for Mark Twain, they all quote him* Thomas Wentworth Higginson, journal entry for Aug. 6, 1878, in *The Letters and Journals of Thomas Wentworth Higginson, 1846–1906*, ed. Mary Thatcher Higginson (Boston: Houghton Mifflin, 1921), 300.

144 *"Big River"* Masago Igawa, letter to the author, July 12, 1995.

144 *animated series* Maria Alejandra Rosarossa, letter to the author, Aug. 31, 1995.

144 *Highly successful productions* Rosarossa, letter to the author.

144 *Films based on Twain's works* Barbara Hochman, letter to the author, July 4, 1995.

144 *an Indian film version* James L. Limbacher, *Haven't I Seen You Somewhere Before? Remakes, Sequels and Series in Motion Pictures and Television, 1896–1978* (Ann Arbor, Mich.: Pierian Press, 1979), 168; and Rasmussen, *Mark Twain A to Z*, 370.

144 *A film version of "The £1,000,000 Bank-Note* Mei Renyi, letter to the author, June 23, 1995.

144 *Twain was first translated into Chinese* Wu Bing, letters to the author, June 29, 1995, and Oct. 10, 1995; Liu Xu-yi, letter to the author, July 5, 1995.

144 *Monteiro Lobato* Carlos Daghlian, letter to the author, June 29, 1995;

144 *Many Japanese* Shunsuki Kamei, "Mark Twain in Japan," *Mark Twain Journal*, 12, no. 1 (Spring 1963):10–11.

144 *self-appointed Ambassador at Large* Mark Twain, quoted in *MC&MT*, 354.

145 *Teachers in the Netherlands, England, Japan, and Portugal* Julia J. G. Muller-Van Santen, letter to the author, July 1, 1995; Stephen Fender, letter to the author, June 30, 1995; Makoto Nagawara, letter to the author, June 14, 1995; Susan Castillo, letter to the author, June 22, 1995.

145 *Dutch teenagers are "riveted by the idea . . .* Muller-Van Santen, letter to the author.

145 *Readers for whom "planned landscapes . . . both physical and social* Stephen Fender, letter to the author.

145 *Readers "who tend to feel cramped under overly rigorous school management"* Nagawara, letter to the author.

145 *A professor in Beijing* Wu Bing, letter to the author, June 29, 1995.

145 *Peter Stoneley's students at Queen's University* Stoneley, letter to the author, June 30, 1995.

145 *In a similar vein, Professor Makoto Nagawara of Kyoto* Nagawara, letter to the author.

145 *Professor Prafulla Kar* Prafulla Kar, letter to the author, June 25, 1995.

146 *anti-imperialism, along with his exposés* Liu Xu-yi, letter to the author; Wu Bing, letter to the author, June 29, 1995.

146 *Twain challenged specifically English traditions* Peter Stoneley, letter to the author.

146 *Twain maintained the same delicate balance* Horst H. Kruse, letter to the author, July 17, 1995; Alfred Hornung, letter to the author, Aug. 15, 1995; Helmbrecht Breinig, letter to the author, July 10, 1995. For more on German responses to Twain, see J. C. B. Kinch, *Mark Twain's German Critical Reception, 1875–1986: An Annotated Bibliography* (New York and Westport, Conn.: Greenwood Press, 1989). See also Mario Materassi, review of Twain's *Gli innocenti all'estero* (ed. Agostino Lombardo [Milan: Lerici, 1961]) in *Atlas* 2, no. 6 (December 1961): 489–92 (translated from *Il Ponte*).

146 *Praise is well, compliment is well* Twain, "Books, Authors and Hats," *Mark Twain's Speeches* (1923 ed.), 343; quoted in *MTAYF*, 5.

147 *legendary animation director* See "Legendary Animation Director Returns to Warner Bros.," Business Wire, Feb. 9, 1994; "Video Roundup," *Arizona Republic*, July 8, 1994, p. D9; Robert Osborne, "Rambling Reporter," Entertainment Newswire, April 8, 1994; "Beep! Beep! Road Runner Lives Again," *Commercial Appeal* (Memphis), Feb. 12, 1994, p. 2C; Jennifer Pendleton, " 'Dogs' Has Its Day . . . ," *Daily Variety*, Oct. 27, 1992, p. 7.

147 *The Coyote—Mark Twain discovered him first* Chuck Jones, *Chuck Amuck: The Life and Times of an Animated Cartoonist* (New York: Farrar, Straus and Giroux, 1989), 34.

147 *Mark my words* Jones, *Chuck Amuck*, 34.

147 *One fateful day* Jones, *Chuck Amuck*, pp. 33–34.

147 *Walt Disney, for example* See Bob Thomas. *Walt Disney: An American Original* (New York: Simon and Schuster, 1976), 36; Richard Schickel, *The Disney Version: The Life, Times, Art and Commerce of Walt Disney*, rev. ed. (New York: Touchstone/Simon and Schuster, 1985), 323.

148 *while devouring Mark Twain's* Roughing It Jones, *Chuck Amuck*, 34.

149 *The coyote is a long, slim, sick and sorry looking skeleton* Twain, *Roughing It*, quoted in Jones, *Chuck Amuck*, 34–35.

149 *are known and accepted throughout the world* Jones, *Chuck Amuck*, 226.

149 *gave me the clue to the speed of the Road Runner* Chuck Jones, quoted in Steven Schneider, *That's All Folks! The Art of Warner Bros. Animation.* (New York: Donald Hunter/Henry Holt,1988), 222.

149 *straightens himself out like a yardstick* Twain, *Roughing It*, 13.

149 *animation* Jones, *Chuck Amuck*, 13.

149 *shortly after World War II* Jones, *Chuck Amuck*, 13.

149 *He dropped his ears* Roughing It, 13.

150 *leave a Bismarck herring* Jones, *Chuck Amuck*, 34.

150 *What grapefruit was to Johnson the cat* Jones, *Chuck Amuck*, 34.

151 *The first Twain impersonation in print* "Un interview cu di Mark Twain," by

Vasile Pop, which appeared in the Dec. 15, 1906, issue of *Luceaufarul* (Morning Star), a "magazine for literature and the arts, appearing twice a month," was not translated into English until 1988, and there is no evidence that Mark Twain was aware of it. See Helen Leath, "An Interview with Mark Twain: A Translation." *Mark Twain Journal* 26, no. 1 (Spring 1988): 25–29.

151 *One wag, writing in 1909* Eugene Angert, asking "Is Mark Twain Dead?," "applied Clemens's analytic methods and reached the conclusion that the writer currently known as Mark Twain was as much an imposter as 'Mark Twain' said Shakespeare was." Angert, quoted in *MC&MT*, 383.

151 *'The Coming of 'Jap Herron'* Emily Grant Hutchings, "The Coming of 'Jap Herron,' " in *Jap Herron* (New York: Michael Kennerly, 1917), 1–42, quote p. 3.

152 *Posthumous "dictations" from Twain by planchette* Eunice Winkler, "The Return of Mark Twain," *Azoth*, 2 (Jan. 1918): 5–17, cited in Leath, "An Interview with Mark Twain," 29. Mildred Burris Swanson, *God Bless U, Daughter* (Independence, Mo.: Midwest Society of Psychic Research, 1968).

152 *Michael Frank, an editor at the Mark Twain project, recalls* Frank related this story to me in a telephone conversation in October 1995.

152 *The picture on the opposite page entitled "The Political Pot"* H. M. and D. C. Partridge, *The Most Remarkable Echo in the World*, illustrated (New York: privately printed [Cosmo Printing Co.], 1933), 21.

153 *This "spurious version"* William M. Gibson, "Introduction," *Mark Twain's Mysterious Stranger Manuscripts*, ed. William M. Gibson, The Mark Twain Papers (Berkeley: University of California Press, 1969), 1–34, quote p. 1. Gibson's introduction remains the most comprehensive analysis of what he calls an "editorial fraud" (1).

153 *to get out another book by "Mark Twain"* Gibson, "Introduction," 4.

155 *that dirty, uncouth, big-nosed* Philip José Farmer, *The Fabulous Riverboat* (New York: Berkeley Books, 1973), 104.

155 *And to think that my old friend* Gore Vidal, *1876: A Novel* (New York: Random House, 1976), 281.

155 *"What is Man?" the aging* Joyce Carol Oates, *A Bloodsmoor Romance* (New York: Warner Books, 1982), 438.

156 *"This is good," I thought* David Carkeet, *I Been There Before* (New York: Harper & Row, 1985), 20.

156 *quite a busy little beehive* Carkeet, *I Been There Before*, 55.

157 *"Hmmm," I though*t Carkeet, *I Been There Before*, 56–57.

157 *He was tired of being the buffoon* Carkeet, *I Been There Before*, 58.

157 *"And what the hell is that?"* Allen Appel, *Twice Upon a Time* (New York: Dell, 1989), 141.

158 *That book was made* Karen Lystra, review of *WHB* in *American Historical Review* v.98 (Dec. 1993): 1559–61.

159 *there warn't really anything the matter* Hucklebery Finn, 2.

159 *got right to work* John Seelye, *The true adventures of Huckleberry Finn as Told by John Seelye*, 2nd ed. with a P. S. by Huck Finn (Urbana: University of Ilinois Press, 1987), x.

159 *So I thought to myself, if the book* Seelye, *The true adventures*, xii.

159 *And now that they've got* their *book*, Seelye, *The True Adventures*, xii.

159 *Jim said that was plain silly* Seelye, *The True Adventures*, xxii. In between writing the original (1970) and the revised (1987) editions of *The True Adventures of Huckleberry Finn*, Seelye wrote a western satire called *The Kid* (1972), which he envisioned as a sequel to *Huckleberry Finn*. (Seelye, conversation with the author, San Diego, May 30, 1996). For another effort to retell Huckleberry Finn through Jim's eyes, see Gerry Brenner's imaginative "More than a Reader's Response: A Letter to 'De Ole True Huck,' " *Journal of Narrative Technique* (1990) rpt. *Mark Twain, "Adventures of Huckleberry Finn": A Case Study in Critical Controversy*, ed. Gerald Graff and James Phelan (Boston: Bedford Books/St. Martin's, 1995), 450–68.

160 *America's foremost symbols of boyhood* William A. Henry III, "Deep Nerve: 'The Boys in Autumn,' by Bernard Sabath," *Time*, May 12, 1986, Theater, p. 98.

160 *A home without a cat* Twain, *Pudd'nhead Wilson*, OMT, 18.

161 *in which the hero or heroine has gone back in the past* Isaac Asimov, "When It Comes to Time Travel, There's No Time Like the Present," *NYT*, Oct. 5, 1986, B1.

161 *"modern slice" of the novel* Anon., "The New Mark Twain," *Northeast* (*Hartford Courant*), Dec. 10, 1989, p. 13.

162 *a convention of the Society for Creative Anachronism* Mike Wavada, "Sir Consultant's Strategic Plan," *Northeast* (*Hartford Courant*), Dec. 10, 1989, p. 16.

162 *When I came to* Joey Fishkin, "A Return to Camelot," *Northeast* (*Hartford Courant*), Dec. 10, 1989, 29.

162 *won the hearts* Anon., "The New Mark Twain," p. 26. *Northeast* (*Hartford Courant*) Dec. 10, 1989, 26.

163 *The mortal Twain* Louis J. Budd, "Impersonators," *MTE*, 390.

163 *Henry Sweets recalls* Henry Sweets, taped interview by author, Hannibal, Mo., June 21, 1995.

163 *In Memphis an assistant director of Parks* "Big Bartlett Event Raises Arts Funds," *Commercial Appeal* (Memphis), Nov. 6, 1995, p. 19A. (The assistant director of Parks and Recreation was Ron Jewell); Mary George Beggs, "Who What Where," *Commercial Appeal* (Memphis), June 9, 1994, p. 2C. (The Shelby County Sheriff's Department staff member was Captain Doug Carver.)

163 *I believe so much in what Mark Twain was trying to do and say* Marvin Cole, quoted in Dena Smith, "College President Leaving Campus to Slip into Role of Mark Twain," *Atlanta Journal and Constitution*, May 1, 1994, p. J4.

163 *one of the best writers in the world* F. X. Brown, quoted in Kathleen Bercaw, "Mark Twain Impersonator Says He Has One of Best Script Writers," *Morning Call* (Allentown, Pa.), Dec. 10, 1995, p. E8.

163 *Ken Richters never planned to "become" Mark Twain* Ken Richters, telephone interview with the author, October 1995.

164 *McAvoy Layne, of Incline, Village* McAvoy Layne, telephone interview with the author, July 1995.

165 *What did he sound like?* Hal Holbrook, "Introduction," *Mark Twain's Speeches*, OMT, xxxi.

165 *I worked in the curve of a baby grand piano* Holbrook, "Introduction," OMT, xxxv.

166 *But when President Eisenhower called out the troops* Holbrook, "Introduction," OMT, xxxvii.

166 *was doing Mark Twain in Oxford, Mississippi* Holbrook quoted in Frederick Waterman, "New England Showcase," UPI, Nov. 26, 1982.

166 *dissent is a tradition* Holbrook, "Introduction," OMT, xxxviii.

167 *The trouble with these beautiful, novel things* Twain, interviewed in *NYT* Dec. 23, 1906; quoted in *Mark Twain Speaks for Himself*, ed. Paul Fatout (West Lafayette, Ind.: Purdue University Press, 1978), 218.

167 *Brad Messerle was not the sort of man* Brad Messerle, interview with the author, Austin, Texas, Nov. 27, 1995.

167 *I took a large room, far up Broadway* Twain, "A Ghost Story," *Sketches, New and Old*, OMT, 215.

168 *This scenario* Information about training IBM's VoiceType Dictation system was obtained from Linda Reda, VoiceType product sales manager for IBM for the northeast region, and from IBM materials which she provided (Linda Reda, telephone interview with the author, October 1995); Dr. David Nahamoo, director of Human Language Technologies at IBM and a member of the team that developed the system and the training program (telephone interview with the author, October, 1995); Ron McGuinn of Houston, who uses and sells the VoiceType® Dictation system (telephone interview with the author, October 1995). See also Cate Corcoran, "Dictation Package Takes Down 70 Words Per Minute," *Infoworld*, Nov. 1993, p. 33; Niklas von Daehne, "Quantum leaps; Communications Technology," *Success*, Sept. 1994, vol. 41, no. 7, p. 31; Florence Olson, "Voice Recognition for the PC Pushes Beyond Old Barriers," *Government Computer News*, Nov. 7, 1994, vol. 13, no. 24, p. 42; Thomas Hoffman, "IBMs Voice System Carves Out User Niche," *Computerworld*, Nov. 21, 1994, p. 40; Nicholas Petreley, "IBM PR Machine Silences Another Terrific Product: VoiceType Dictation," *InfoWorld*, May 1, 1995, p. 101; Wayne Rash, Jr., "IBM Personal Dictation System; IBM Continuous Speech System; IBM Corp; Software Review; One of Five Evaluations of Voice-recognition Systems in "Talk Show," *PC User*, May 18, 1994, vol. 13, no. 22, p. 213; Steve Boxer, "IBM's Replacement for the Keyboard: Voice Recognition Systems; IBM Personal Dictation System . . .," *PC User*, May 18, 1994, no. 235, p. 40; Anon., "IBM's Most Powerful Speech-Recognition Technology on the Desktop: State-of-the-Art Dictation System Available and Affordable," Business Wire, Nov. 2, 1993.

169 *an Improvement in Type-setting Machines* MC&MT, 283.

169 *The patent he requested* MC&MT, 285.

169 *It took the patent office. . . .* Samuel Charles Webster, ed., *Mark Twain, Business Man* (Boston: Little, Brown, 1946), 311.

170 *When it was working properly* MC&MT, 285.

170 *At 12:20 this afternoon a line of movable types* Samuel L. Clemens to Orion Clemens, Jan. 3, 1889 (Hartford), in *MTL*, 2:506–8.

170 *one of the hundred and one devices and inventions* Samuel L. Clemens, rejected version of Typothetae Speech (Jan. 1886), in *Mark Twain Speaking*, ed. Paul Fatout, 205.

172 *a most great and genuine poet* Samuel L. Clemens, quoted in Sherwood Cummings, "Science," *MTE*, 665. For more on Twain's attitudes toward and awareness of science and technology, see Cummings's magisterial *Mark Twain and Science: Adventures of a Mind* (Baton Rouge: Louisiana State University Press, 1988).

173 *the creators of this world* A Connecticut Yankee, 323.

173 *He tried out* Cummings, "Science," *MTE*, 665.

173 *And he was intrigued by* The key source for Twain's interest in Tesla's experiments is Margaret Cheney, *Tesla: Man Out of Time* (Englewood Cliffs, N.J.: Prentice Hall, 1981), which includes references to Twain—based on Twain's letters to Tesla in the Library of Congress—throughout the volume. Twain's own letters are the source of most of the information about his own inventions and experiments, a subject which his biographers, from Paine to Kaplan, have treated in detail. Samuel Charles Webster, in *Mark Twain, Business Man*, has addressed the financial dimensions of Twain's investments in technology.

173 *Dear Charley* Samuel L. Clemens to Charles Webster (October 1884, undated letter), in Webster, *Mark Twain, Business Man*, 280

174 *first person in the world* Clemens, quoted in Cummings, "Science," *MTE*, 665.

174 *I DON'T KNOW WHETHER* Clemens to William Dean Howells, Dec. 9, 1874 (Hartford), in *MTL* 1:238. Paine reproduces this letter using uppercase and lowercase letters, although Clemens used only uppercase.

174 *I AM TRYING TO GET THE HANG OF THIS* Clemens to Orion Clemens, Dec. 9, 1874 (Hartford), facsimile of letter reproduced in *Paine*, 1, insert between pp. 536 and 537.

175 THE MACHINE HAS SEVERAL VIRTUES Clemens to Orion Clemens, Dec. 9, 1874, in *Paine*, 1, insert between pp. 536 and 537.

175 *BECAUSE I AM FULL OF RHEUMATISM* Samuel Clemens to Mr. and Mrs. Gerhardt, Feb. 24, 1882 (Hartford). Mark Twain's previously unpublished words quoted here are © 1997 by Chemical Bank, Trustee of the Mark Twain Foundation which reserves all reproductive or dramatization rights in every medium. Quotation is made with the permission of the University of California Press and Robert H. Hirst, General Editor of the Mark Twain Project. Each subsequent quotation of previously unpublished words by Mark Twain is also © 1997 and is identified by an asterisk (*) in its citation. This letter is in the Mark Twain Collection of the James S. Copley Library, La Jolla, Ca, and is quoted with permission.

175 *Twain's enthusiasm for technological innovation* See the following for more on articles and products mentioned in this paragraph: Charles Petzold, "The Marriage of Text and Graphics," *PC Magazine*, Oct. 31, 1989, vol. 8, no. 18, p.

337; Gordon McComb, *WordPerfect 5.1 Macros and Templates* (New York: Bantam Books, 1990), vii; Work-in-progress demonstration of "Nile: Passage to Egypt" CD-ROM produced by Discovery Channel and Human Code at Author's Guild Southwest Conference, Austin, Texas, Jan. 17, 1995; Harley Jebens, David Plotnikoff, and Andy Wickstrom, "Powerhouse: Hardware, Software," *Austin American-Statesman*, July 20, 1995, vol. 40, p. 26; "Twain's World" © 1993 Bureau Development, Inc.; Vincent Canby, "Screen: Adventures of Mark Twain," *NYT*, Jan. 17, 1986, p. C8; Sheila Benson, "Movie Review," *LAT*, Jan. 17, 1986, sec. 6, p. 21; "Robert Blau, " 'Twain' Gummed Up By Wrongheadedness," *Chicago Tribune*, Apr. 21, 1986, Tempo, p. 7; Brian Lowry, "Clay People Invade World of Advertising: Vinton Shop Goes 'Low-Tech,' " *Advertising Age*, Oct. 20, 1986, p. 110.

176 (figure) *YOU SEE I AM WRITING THAT WAY AGAIN** Samuel L. Clemens to Mr. and Mrs. Gerhardt (Hartford), Mar. 23, 1882. Mark Twain Collection of the James S. Copley Library, La Jolla, Ca.; quoted with permission.

177 *Technology Without Any Interesting Name* Sean Cortez-Mathis of Adobe Systems, telephone conversation with the author, November 1995. (Adobe Systems acquired Aldus, a firm that played a key role in developing the TWAIN scanning standard. According to Mathis, when researchers trying to come up with a name for the scanning standard found that everything they thought of was already taken, they finally settled on <u>T</u>echnology <u>W</u>ithout <u>A</u>ny <u>I</u>nteresting <u>N</u>ame).

178 *a whole bunch of folks found out* script quoted in Michael Wallace, "Mickey Mouse History: Portraying the Past at Disney World," in *History Museums in the United States: A Critical Asssessment*, ed. Warren Leon and Roy Rosenzweig (Urbana: University of Illinois Press, 1989), 174–75.

178 *so life-like it's downright eerie* David Harding, "Butter Sculpture a Hit at Fair," UPI, August 11, 1989, Regional News.

178 *the robot is the size of a normal man* Harding, "Butter Sculpture."

178 *"Virtual Mark Twain"* MacAvoy Layne and Gary Jesch, telephone interviews with the author, July 1995.

178 *a digital library of Mark Twain's works* Gary Jesch, telephone interview.

178 *amounted to a passion* Paine, as quoted in Cummings, "Science," 664.

178 *As a citizen of his time* Cummings, "Science," 664.

179 *the machine becomes a means of death* Jerry Thomason, "Technology," *MTE*, 728.

179 *telephonic conversation* Twain, "The Loves of Alonzo Fitz Clarence and Rosannah Ethelton," originally published in *The Atlantic Monthly*, Mar. 1878.

180 *The trouble with these beautiful, novel things* Twain, *NYT*, Dec. 23, 1906; quoted in *Mark Twain Speaks for Himself*, 218.

180 *Yet some of his mad schemes* Bobbie Ann Mason, "Introduction," *The American Claimant*, OMT, xxxvi. Cynthia Ozick, "Introduction," Mark Twain, *The Man That Corrupted Hadleyburg and Other Stories and Essays*, OMT, xliv; Malcolm Bradbury, "Introduction," *The £1,000,000 Bank-Note and Other New Stories*, OMT, xliii.

180 *without doubt the most stupendous* Twain, *The Autobiography of Mark Twain*, ed. Charles Neider (New York: Harper, 1959), p. 302.

181 *Disneyland's* Mark Twain *riverboat* "Disneyland's Mark Twain ASCII Animation," posted by Don Bertino on ASCII Art USENET Group.

181 *And he might puzzle over* The prankster's obscene sketch of Uncle Silas that delayed the publication of the first American edition of *Huckleberry Finn* is included in "Mark Twain's *Huckleberry Finn*: Text, Illustrations, and Early Reviews," a highly acclaimed exhibit by Virginia H. Cope of the University of Virginia, which may be accessed through Jim Zwick's "Mark Twain Resources on the World Wide Web."

181 *Jim Zwick's phenomenal* "Mark Twain Resources on the World Wide Web" was selected as one of the "Best 1,001 Internet Sites" by *PC Computing*, and and appears on WebCrawler Select's list of the "Best of the Net." Created and maintained by Jim Zwick, editor of the acclaimed book *Mark Twain's Weapons of Satire: Anti-Imperialist Writings on the Philippine–American War* (Syracuse, N.Y.: Syracuse University Press, 1992), this elegant, clearly organized and constantly updated web site links readers to a dizzying range of Twain related material.

181 *the Mark Twain Forum* "There's almost a sense that you're chatting about a distant relative: someone else in family will remember some detail you forgot," says Taylor Roberts, the M.I.T. linguist who founded the Mark Twain Forum in 1992. Living in Toronto at the time, Roberts had a strong interest in Mark Twain and no one to share it with. Having gotten on e-mail himself the year before, Roberts started the electronic-mail discussion group with a handful of people in the United States who had e-mail accounts and an interest in Twain. Within four years the Mark Twain Forum had hundreds of subscribers from across the United States, as well as Australia, Austria, Canada, Germany, Greece, India, Israel, Japan, Saudi Arabia, and the United Kingdom. Exchanges on the forum range from purely informational to highly subjective. Scholars often use the forum to post queries and findings central to new research in the field.

EPILOGUE

184 *a fantasy of adolescence* Julius Lester, "Morality and *Adventures of Huckleberry Finn*, in *Satire or Evasion?*, ed. Leonard, Tenney, and Davis 199–207.

184 *Oh my God. Where did he write that?* Ralph Ellison, taped interview by the author.

185 *an "American" writer in the hemispheric sense* Maria Alejandra Rosarossa, letter to the author, Aug. 31, 1995.

185 *Similarly, political scientist* Ralph Buultjens, interview with the author, March 1993, Cambridge, Eng. (Clare Hall), and telephone conversation with the author, May 1996.

185 *a very offensive specimen* quoted in Dennis Welland, *Mark Twain in England* (London: Chatto & Windus, 1978), 36.

185 *one of the greatest living geniuses* Andrew Lang, "International Girlishness,"

Murray's Magazine (London) 4 (October 1888): 433–41; quoted in Thomas Asa Tenney, *Mark Twain: A Reference Guide* (Boston: G. K. Hall, 1977), 16.

185 *In the twentieth century* Tony Tanner, *Reign of Wonder* (Cambridge: Cambridge University Press, 1965); Peter Messent, "Afterword," *The American Claimant*, OMT; Messent, *New Readings of the American Novel* (London and New York: Macmillan, 1987), 204–42; Messent, *Mark Twain* (forthcoming); Peter Stoneley, *Mark Twain and the Feminine Aesthetic* (Cambridge: Cambridge University Press, 1992). Despite the occasional ruffled feather or moment of bewilderment, "the evidence of Mark Twain's popularity with the English reader is incontrovertible" (Welland, *Mark Twain in England*, 32). See also Baetzhold, *Mark Twain and John Bull*.

185 *You don't ever seem to want to do* Twain, *Huckleberry Finn*, 300–301.

185 *a person who does things because* Twain, *Mark Twain's Notebook* (1935 ed.), 169; quoted in *MTAYF*, 7.

186 *What is it that confers* The Innocents Abroad, OMT, 266.

187 *had something which he thought would overcome us* Twain, *The Innocents Abroad*, OMT, 292.

187 *Guides cannot master*, The Innocents Abroad, OMT. 293.

187 *don't take no stock*, Huckleberry Finn, 2.

187 *the part wherein Huck* Grant Wood, "My Debt to Mark Twain," *Mark Twain Quarterly* 2 (Fall 1938): 6, 14, 24.

188 *plain style assaulted* Bobbie Ann Mason, "Introduction," *The American Claimant*, OMT, xxxiv, xxxiii.

188 *the state in which Twain's parents were raised* Twain's mother was born and grew up there, and his father moved there at age seven or eight.

188 *"warned against the threat to "proper English"* Mordecai Richler, "Introduction," *The Innocents Abroad*, OMT, xlv.

189 *Indeed, Twain is cited* This tally of Twain citations in the current *OED* comes from Taylor Roberts. (Bruce Michaelson, *Mark Twain on the Loose* [Amherst: University of Massachusetts Press, 1995], 2.)

189 *Ben Okri's beautifully written* Cited on second page of Okri, *The Famished Road* (New York: Vintage, 1992).

190 *The great Argentine writer* Henry Sweets, interview by the author. For Yevtushenko's interest in Twain, see Mary Wade Burnside, "Well-Versed Poet Yevgeny Yevtushenko's Long, Strange Trip from Russia to Tulsa," *Charleston Gazette*, March 28, 1996, p. 1D.

190 *pedagogical blunder* Josef Skvorecky, "Huckleberry Finn: Or, Something Exotic in Czechoslovakia," *NYT*, Nov. 8, 1987, Book Review, p. 47.

190 *It seems unlikely that* John Nathan, "Introduction," *Teach Us To Outgrow Our Madness: Four Short Novels by Kenzaboro Ōe*, trans. John Nathan (New York: Grove Press, 1977), xii. Ōe made these comments about Twain during conversations in English with Nathan that took place several mornings a week for three months in 1964.

190 *The heroes of Ōe's fiction* Nathan, "Introduction," xiii.

190 *a devoted fan of Twain's book* Susan J. Napier, *Escape from the Wasteland: Romanticism and Realism in the Fiction of Mishima Yukio and Ōe Kenzaburo*

(Cambridge, Mass.: Council on East Asian Studies, Harvard University, Harvard University Press, 1991), 32. Napier cites Ōe Kenzaburo, "Hakuruberi Finu to hiro no mondai," *Ōe Kenzaburo dōjidiraionshū*, vol. 7 (Tokyo: Iwanami Shoten, 1981), 40–62.

191 *a watershed event* Frederik Pohl, "Introduction," *Captain Stormfied's Visit to Heaven*, OMT, xxxii. Pohl has won most of the awards the science-fiction field has to offer, including the "Grand Master" Nebula award for lifetime contributions in the field.

191 *the first time I ever read the story* Ursula Le Guin, "Introduction," *The Diaries of Adam and Eve*, xxxv. Le Guin went on to win a National Book Award, five Hugos, and four Nebula awards.

191 *was magic to me* Willie Morris, "Introduction," *Life on the Mississippi*, OMT, xxxi.

191 *It was A Connecticut Yankee* Arthur Miller, "Introduction," *Chapters From My Autobiography* , OMT, xxxi.

191 *Toni Morrison returned* Toni Morrison, conversation with the author, December 1991, and Morrison, Introduction," *Adventures of Huckleberry Finn* , OMT, xxxiii.

191 *Twain as an important forebear* See, for example, *WHB*, 4,9, 137–40; Fishkin, "Mark Twain and Women," in *CCMT* 69,73; Tom Quirk, *Coming to Grips with Huckleberry Finn* (Columbia: University of Missouri Press, 1993), 106–46;

191 *to be an artist* David Bradley, "Introduction," *How to Tell A Story*, OMT, li.

192 *The difference between the* right *word* Mark Twain, "Reply to the Editor of 'The Art of Authorship' " (1890) *CTSSE, 1851–1890*, 946.

192 *I thought about James T. Stewart* Bradley, "Introduction," OMT, lviii–lix.

193 *perhaps the most significant work* Art Seidenbaum, "Superior Saga Tracks Heritage from Slave Times," *LAT*, April 8, 1981, sec. 5, p. 2; Larry Swindell, "Slavery Is Still the Black Man's Burden," *Fort Worth Star Telegram*, April 19, 1981, p. 6F. The Chaneysville Incident won the PEN/Faulkner prize in 1982, as well as an Academy Award from the American Academy and Institute of Arts and Letters.

195 *There have been daring people* Twain, "Fenimore Cooper's Literary Offenses," in *How to Tell A Story* , OMT, 114–116.

196 *I closed the cover stunned* Jane Smiley, "Say It Ain't So, Huck. Second Thoughts on Mark Twain's 'Masterpiece.' " *Harper's Magazine* (Jan. 1996), 61.

196 *It was at that momen*t Ralph Wiley, *Dark Witness: When Black People Should Be Sacrificed (Again)* (New York: One World/Ballantine Books, 1996), 22.

196 *the end of* Huckleberry Finn Ralph Wiley, *Dark Witness*, 51.

197 *a little worried* Julia Rosenbloom, telephone interview by the author, Dec. 1994 and Apr. 1996.

197 *Mark Twain was being quite realistic* Ralph Ellison, interview by the author.

197 *The matchless lessons* Dean Matthew S. Santirocco, telephone interview by the author, December 1995. During freshman orientation, students heard Prof. Cyrus Patell lecture on Twain's novel and participated in small-group discussions of *Pudd'nhead Wilson* led by Prof. Pam Shermeister and other

faculty members. They heard biology professor David A. Scicchitano speak on "The Human Genetic Fingerprint" and heard literature Professor Phillip Harper lecture on identity issues more generally. Harper feels he may have heard resonances of the discussions in which *Pudd'nhead Wilson* was the centerpiece in his undergraduate course that term: "The students in that class, even those I know to be freshmen, seem to have an understanding of the complexity of the issues related to identity that I found quite striking. [They showed] a more sophisticated understanding than students I'd been teaching at Harvard." (Harper, Patell, Shermeister: telephone interviews by the author, Dec. 1995 and Jan. 1996).

197 *the Negro [had] been pushed* Ralph Ellison, "Twentieth-Century Fiction and the Black Mask of Humanity," in Ellison, *Shadow and Act* (New York: New American Library, 1964), 51.

198 *collapse of civil rights* Toni Morrison, "Introduction," *Adventures of Huckleberry Finn*, OMT, xxxvi. For detailed citations for this thread in scholarship, see *WHB*, 74–76, 219–47. See also Butcher, "Mark Twain's Installment on the National Debt"; and Messent, *Mark Twain*, forthcoming.

198 *the old master class* Douglass, quoted in anon., "FRED DOUGLASS! His Great Speech Yesterday," *EA*, Aug. 4, 1880, p. 2.

198 *his revisions of manuscript page 751* Victor A. Doyno, "Textual Addendum," Mark Twain, *Adventures of Huckleberry Finn* (New York: Random House, 1996), 382; a facsimile of the corrected manuscript page appears on p. 415.

199 *It was that one part black* Mark Twain, quoted in William Dean Howells, *My Mark Twain*, 35–36.

199 *contrast between our ideals and activities* Ralph Ellison, interview by the author.

200 *what white racists call black Ph.D.s* Malcolm X, *Autobiography of Malcolm X*, 284.

200 *even the best educated negro is at a disadvantage* Mark Twain, ["The Man With Negro Blood"], reprinted and discussed in Shelley Fisher Fishkin, "False Starts, Fragments and Fumbles: Mark Twain's Unpublished Writing on Race," *Essays in Arts and Sciences* 20 (October 1991): 17–31.

201 *the silent assertion* Twain, "My First Lie and How I Got Out of It," *CTSSE*, 1891–1910, 440.

201 *the town's leading white citizens* Ralph Ellison, *Invisible Man* (New York: Vintage Books, 1989), 17.

201 *battle royal* Ellison, *Invisible Man*, 17–29.

201 *If the ideal of achieving* Ellison, "Introduction," *Invisible Man*, xx.

201 *a novel could be fashioned* Ellison, "Introduction," *Invisible Man*, xx–xxi.

202 *to bury the combustible issues* Morrison, "Introduction," OMT, xxxvi.

202 *the power of* Uncle Tom's Cabin Smiley, "Say It Ain't So, Huck," 65.

202 *One has to look* Ellison, interview by the author.

202 *Irony was not for those people Pudd'nhead Wilson*, OMT, 71.

202 *A lot of snotty academics* David Bradley, "Why Teach *Huckleberry Finn*?," speech presented at Drake University, Des Moines, Iowa, October 1995.

ACKNOWLEDGMENTS

Many people helped make this book happen, and it is a pleasure to acknowledge a number of them here.

The following read this book in its entirety in manuscript form: Kevin Bochynski, David Bradley, Louis J. Budd, Victor A. Doyno, Carol Fisher, Milton Fisher, Jim Fishkin, Joey Fishkin, T. Walter Herbert, Elizabeth Maguire, Peter Messent, Carla Peterson, R. Kent Rasmussen, Lillian Robinson, Jeffrey Rubin-Dorsky, and Siva Vaidhyanathan. They did much more than offer invaluable suggestions, corrections, and comments. In addition to providing me with a stream of out-of-print novels and fugitive references, Kevin Bochynski transformed an old computer of mine into a magic machine containing a library of Twain's works in searchable electronic form. David Bradley helped give me the courage to look at history without flinching and taught me how to find the shape of my story. Lou Budd and Vic Doyno brought their matchless knowledge of Mark Twain to bear on the project, and generously shared obscure facts and little-known paper trails. Milton and Carol Fisher helped shape this book from the start. Jim Fishkin and Joey Fishkin propelled me to the finish line. Answering Walt Herbert's hard questions helped clarify my thinking on key issues. Elizabeth Maguire persuaded me to drop everything and write this book. Peter Messent, Carla Peterson, Lillian Robinson, Jeffrey Rubin-Dorsky and Siva Vaidhyanathan encouraged me at crucial times and stimulated my thinking in important ways. R. Kent Rasmussen was marvelously "on call" to answer endless questions about a dizzyingly diverse range of subjects, including how to structure the book.

I am grateful to Oxford University Press senior vice president and trade publisher Laura Brown for believing in this book and its author. I appreciate the support, as well, of Oxford president Edward Barry and vice president and editorial director Helen McInnis. Literature editor T. Susan Chang did a superb job of helping me shape my material; her meticulous attention and excellent judgment were crucial to bringing this project to fruition. Joy Johannessen's genius at helping writers say what they really meant to say is legendary; it was a privilege and pleasure to have her edit this book. Invaluable assistance was also provided by project editor Kathy Kuhtz Campbell, trade editorial, design and production manager Adam Bohannon, art director David Tran, assistant editor Elda Rotor, editorial assistant Rahul Mehta, marketing director Amy Roberts, publicity director Susan Rotermund, and the entire Oxford staff.

I would like to thank the following people for having agreed to be interviewed for this book: Martha Adrian, Nancy A. Douglas Ash, Ruth Baker, Donald Blandford, Betsy Bleigh, Faye Bleigh, Jane Bleigh, Joann Bringer, Ralph Buultjens, Jocelyn Chadwick-Joshua, Howard Coleman, Sean Cortez-Mathis, Hiawatha Crow, Grant Ewart, Reverend Anne Facen, Martha Featherman, Dixie M. Forte, Estel Griggsby, Phillip Harper, James Horton, Dan Hurley, Helen Jerome, Robert Jerome, Gary Jesch, McAvoy Layne, Ron McGuinn, Tyrone Marsh, Brad Messerle, David Nahamoo, Cyrus Patell, Linda Reda, Ken Richters, Taylor Roberts, Julia Rosenbloom, Matthew Santirocco, Pam Schirmeister, Mayor Richard Schwartz, Henry Sweets III, Reverend Robert Tabscott, John Vlach, and Senator Royce West.

I am grateful to the following teachers and scholars from Argentina, Brazil, China, France, Germany, Greece, India, Italy, Israel, Japan, the Netherlands, Poland, Portugal, Saudi Arabia, Switzerland, Turkey, and the United Kingdom for their illuminating correspondence or conversations about responses to Mark Twain outside the United States: Isul Bas, Marilla Battilana, Sabine Bauman, Hans Bertens, Wu Bing, Jean-Marie Bonnet, Guido Carboni, Susan Castillo, Carlos Daghlian, Theo D'haen, Jerzy Durczak, Stephen Fender, Katia Georgoudaki, Shoji Goto, Wolf Harranth, Barbara Hochman, Alfred Hornung, Hiroyoshi Ichikawa, Masago Igawa, Iwao Iwamoto, Chika Kaise, Yoshio Kanaya, Prafulla Kar, Holger Kersten, Horst Kruse, A.R. Kutrieh, Albert Locher-Bärtschi, Mario Materassi, Peter Messent, Kurt Müller, Julia J.G. Muller-Van Santen, Makoto Nagawara, Yorimasa Nasu, Gigliola Nocera, Masako Notoji, Hiroshi Okubo, Gordon Poole, Gert Raeithel, Mei Renyi, Maria Alejandra Rosarossa, Viola Sachs, Dietmar Schloss, Peter Stoneley, Asli Tekinay, Ralph Willett, and Liu Xu-yi. I am also grateful to the U.S.I.S. for having given me the opportunity to lecture on Mark Twain at foreign universities.

I appreciate the cheerful support that graduate coordinator Melanie Livingston, administrative assistant Janice Bradley, and office assistant Gilbert Porter of the American Studies Program of the University of Texas always graciously provided. The undergraduates and graduate students in my classes at Yale and at the University of Texas over the years made immeasurable contributions to my understanding of Twain and his role in the classroom.

I am indebted to the Mark Twain House in Hartford for many things, not least of which is the string of extraordinary fall symposiums and summer teachers institutes to which they have invited me over the years, and which have played such an important role in my education. In addition, Executive Director John Boyer, Associate Director and Education Director Debra Petke, curator Marianne Curling, assistant director of education Britt Gustafson, and administrative assistant Pam Collins shared newspaper articles, television transcripts, advertisements and other materials with me, and were always available to answer my questions. Photography curator Beverly Zell did a marvelous job of tracking down and supervising the reproduction of many of the photos included in this book.

I also appreciate the assistance provided by Robert H. Hirst, general editor of the Mark Twain Project of The Bancroft Library in Berkeley, associate editors Michael Frank and Lin Salamo, graduate intern Ann Bui, and administrative assistant Brenda J. Bailey. Like all Twain scholars, I am deeply indebted to the Mark Twain Project for the important ongoing publications and research that play such a central role in our work.

Carol Beales and Ron Vanderhye of the James S. Copley Library in La Jolla went out of their way to make my research trip to their institution enjoyable and productive. I am indebted to Henry Sweets III, Director of the Mark Twain Boyhood Home in Hannibal, and to Hannibal historians Hurley and Roberta Hagood for having shared their time and knowledge with me as generously as they did. I am grateful to Jervis Langdon Jr. for having made Quarry Farm a haven for scholars, and to Gretchen Sharlow, Director of the Elmira College Center for Mark Twain Studies, for having invited me there. I appreciate the courtesies extended to me in Elmira by Michael Kiskis and Karen Ernhout, and by Mark Woodhouse, archivist of the Gannett-Tripp Library at Elmira College. I would also like to thank the staffs of the Mark Twain Boyhood Home, the Hannibal Public Library, the Elmira Public Library, the Missouri State Historical Society, the Texas State Historical Society, the Chemung County Historical Society, the Yale University Archives, and the University of Texas's Perry-Castaneda Library, Harry Ransom Humanities Research Center and Fine Arts Library for the kind help they gave me.

Many people generously provided me with materials that had bearing on my research. JoDee Benussi was kind enough to send me a copy of what is probably the most obscure and bizarre book written about Twain, *The Most Remarkable Echo in the World*. Donald Blandford, Jocelyn Chadwick-Joshua, Marvin Cole, James Collins, Garrett Condon, Hiawatha Crow, Carol Plaine Fisher, Fannie Fishkin, Michael Frank, Gennie Gordon, Hurley and Roberta Hagood, Alice Jaffe, McAvoy Layne, Gretchen Sharlow, Charles Srebnik, and Reverend Robert Tabscott shared newspaper clippings and photographs. Pegge Bochynski and Ruth Perry sent books. Joyce Cohen, Bobby Fishkin, Gwen Frankfeldt, Hal Holbrook, Megan Williams, Phyllis Gillis, Helen Jerome, Cheryl Malone, Glen Morrow, Susan Napier, Barbara Schmidt, Laura Schwartz, Henry Wonham, and Jim Zwick provided me with newsletters, magazines, journals, advertisements, personal correspondence, historical documents, photocopies of reference sources, and other materials. Margaret Osborne and Susan Yecies were particularly supportive during the sometimes manic moments my research engendered. Jeff Meikle, Peter Messent, Tom Perry, Gretchen Sharlow, and Bruce Wilson shared their unpublished research with me. John Marszalek and Michael Kassel graciously answered my questions. Gerald Hauck helped me out with some German translations. Conversations with Robert Crunden, Gerald Early, James Horton, John Stephens, and William Zinsser pointed me towards key people and sources. Gregg Camfield, Joel Dinerstein, Gretchen Sharlow, and Max Stinchcomb read early portions of the manuscript, and provided helpful responses.

I am grateful to the writers and scholars who contributed to the Oxford Mark Twain, whose essays I edited as I wrote this book, for having added immeasurably to my understanding of Mark Twain and his legacies: Russell Banks, Anne Bernays, Roy Blount Jr., Malcolm Bradbury, David Bradley, Frederick Busch, E.L. Doctorow, Nat Hentoff, Hal Holbrook, Charles Johnson, Erica Jong, Ward Just, Justin Kaplan, Ursula Le Guin, Judith Martin, Bobbie Ann Mason, Arthur Miller, Willie Morris, Toni Morrison, Walter Mosley, Cynthia Ozick, George Plimpton, Frederik Pohl, Mordecai Richler, Lee Smith, Gore Vidal, Kurt Vonnegut Jr., Sherley Anne Williams, and Garry Wills, along with David Barrow, Richard Bucci, Louis J. Budd, Gregg Camfield, Pascal Covici, Jr., Sherwood Cummings, Beverly David, Vic Doyno, Everett Emerson, Leslie Fiedler, Susan K. Harris, Hamlin Hill, Lawrence Howe, M. Thomas Inge, Fred Kaplan, Michael Kiskis, Judith Yaross Lee, James S. Leonard, Peter Messent, James A. Miller, Forrest G. Robinson, Lillian S. Robinson, Jeffrey Rubin-Dorsky, Ray Sapirstein, Laura E. Skandera-Trombley, David E.E. Sloane, David Lionel Smith, Albert Stone, Linda Wagner-Martin, James D. Wilson, and Henry B. Wonham.

I would like to thank the Mark Twain Foundation and the Mark Twain Project for permission to publish Mark Twain's unpublished words. I am grateful to Helen Copley for permission to publish an unpublished letter from Clemens in the Mark Twain Collection of the James S. Copley Library. I would also like to thank the following for the photographs they supplied and the permission to publish them: R.W.B. Lewis and Nancy Lewis, for permisson to use the photograph of Ralph Ellison from their personal collection; Chuck Jones and Melanie Behnke of Chuck Jones Enterprises for the right to reproduce Jones' sketch of Mark Twain; Paul Rodriguez, and Kelly Nielson of Prism Entertainment for the photo from *A Million to Juan*; William Lynch of Guinness Import Co., for arranging permission to publish the Bass Ale Poster; Hal Holbrook, Bobbie Ann Mason and Frederik Pohl for having graciously supplied personal photos; Random House for the photo of Toni Morrison, Grove Press for the photo of Kenzaburo Oē, Vintage Press for the photo of Ben Okri; R. Kent Rasmussen for his kindness in letting me publish photos he took in Hannibal; Carole Patterson for permission to publish her photo of David Bradley; the University of Maryland at College Park Library for permission to publish their photo of Mary Ann Cord; and the Mark Twain House, for photos of Mark Twain, of the 1935 Mark Twain Centennial, Jervis Langdon, Quarry Farm, the Twain house, the 1870s advertisement and the Paige compositor.

This book would not have been possible without the support of my family, most of whose names have already appeared here, as sources of valued editorial advice and useful research materials. I would also like to thank David Fishkin, Gillis, Leonard and Moss Plaine, Jorge Radovich and Milded Hope Witkin for their encouragement.

The intensity with which I immersed myself in this project required Herculean patience on the part of the people I live with. My husband and my sons did more than put up with me. They assembled bookcases, fixed computer glitches, took phone messages, marked up drafts, listened and voiced opinions, and cheered me on all along the way. I couldn't have done it without them.

INDEX

Page numbers in italics refer to photographs and illustrations.

A.M.E. Church, 46
Abolitionism, 53, 55, 74, 75, 79–81, 95, 197, 208, 221. *See also* Slaves and slavery; Underground Railroad
Adaptations. *See* Dramatizations and adaptations of Twain's works
Adrian, Martha, 32–33
The Adventures of Huck Finn (film), 143
Adventures of Huckleberry Finn (Twain), 7, 13, 14, 28, 31, 32, 44, 51, 53, 64, 65, 77, 98, 128, 160–61, 172, 175, 181, 184, 194–98, 230, 246; as antiracist novel, 4–6, 17, 23, 36, 52, 61, 62, 89, 93, 94, 97, 100–101, 106–108, 123, 197, 199, 202–203; attacked as racist, 100–101, 107–108, 114–16, 121, 133, 154–55, 196, 228; in the classroom, 4–6, 8, 9, 33, 107, 114–24, 145, 185, 196–97, 207–8; as commentary on race relations in post-reconstruction era, 97, 119, 197–203, 227, 231; dramatizations and adaptations of, 39–40, 100, 139, 143–44, 166, 239; ending of, 33–34, 66, 97, 118, 119–21, 137, 198–200, 202–203, 231; genesis of, 21, 23–24, 58–61, 89, 93–94, 97, 110–13, 117; influence of,

on other writers, 26–27, 154, 189–93; sequels to, 159–60; slavery in, 5, 17, 56, 58–59, 61, 63, 66, 89, 93–94, 121, 123, 190, 192, 196–97; translations of, 144. *See also* Finn, Huckleberry (fictional character); Finn, Pap (fictional character); Jim (fictional character); Sawyer, Tom (fictional character)
The Adventures of Tom Sawyer (film), 92
The Adventures of Tom Sawyer (Twain), 4, 13, 14, 17, 30–32, 36, 42, 44, 45, 54, 61, 64, 65, 91–94, 137–38, 225, 239; in the classroom, 33; dramatizations and adaptations of, 39–40, 92, 139, 142–43, 144; slavery in, 92–93; translations of, 144. *See also* Finn, Huckleberry (fictional character); Injun Joe (fictional character); Sawyer, Tom (fictional character)
Advertising, 9, 18, 64, 128–31, *130*, *132*, 209
African-American history and culture, 4–6, 13–25, 28–29, 34–36, 40–57, 61–62, 70–75, 80–90, 95–99, 103–13, 116–23, 191–203, 227, 231; and education, 15–17, 34–37, 40, 41, 46, 50, 67–69, 80–81, 208, 212–13; invisibility and erasure of, 16–17, 39–40, 49, 51, 61, 62, 68, 121–22, 216;

African-American history and culture (*continued*)
language use and rhetorical traditions, 22, 29,
51, 85–89, 87, 88, 109–13, 111–13, 230; and
museums, 34, 51, 114, 216. *See also* Elmira,
N.Y.; Free blacks in the slave South; Hannibal,
Mo.; Race relations; Racism; St. Louis, Mo.;
Slaves and slavery; Underground Railroad; and
specific African Americans
African-American Museums, 34, 114, 216
African Methodist Episcopal Church. *See* A.M.E.
Church
American Citizen, 103, 104, 105
The American Claimant (Twain), 175, 180
American Places (Zinsser), 26
Anderson, Maxwell, 139
Anderson, Sherwood, 189
Androids, 177–78
Angert, Eugene, 241
Animation: animated cartoons, 141, 144, 146–50;
animatronics, 177–78; Claymation, 175; Virtual
Mark Twain, 178
"An Ante-Bellum Sermon" (Dunbar), 119
Appel, Allen, 157–58
Arac, Jonathan, 194–95
Arciola, Anthony, 207
Argentine responses to Twain, 8, 26–27, 144, 185,
189, 207
Armstrong, John, 19, 209
Arthur, Chester, 73
Asimov, Isaac, 154, 161, 190
Assault at West Point (film), 218
Assault at West Point (Marszalek), 218
Asselineau, Roger, 146
Australia, 107, 144
Austria, 144
Autobiography of Malcolm X, 116
"The Awful German Language" (Twain), 139,
146, 238
Ayres, John, 91

"Bad faith," 14, 60, 61
Baker, Ruth, 40–41, 213
Baldwin, James, 90, 193
Banks, Russell, 189
Baptist church, 15, 38. *See also* Southern Baptist
Convention
"Barefoot Children" (Buffett), 139
Beecher, Rev. Thomas, 221
Bell, Alexander Graham, 173
Bellow, Saul, 191
Best Man Wins, 138, 237
Bierce, Ambrose, 135–36
Big River, 143, 144
Bing, Wu, 207, 239, 240
Black Betty (Mosley), 128
Black John (slave), 51
Black Reconstruction in America (Du Bois), 119
Blandford, Alsace, 84, 223, 224
Blandford, Donald, 84, 89, 90, 91, 223, 224
Blandford, Thomas, 84–85
Blankenship, Benson/Bence, 21, 63
Blankenship, Tom, 51, 64

Bleigh, Faye, 33–35
Blockson, Charles L., 216
A Bloodsmoor Romance (Oates), 155
Bonanza, 10
Borges, Jorge Luis, 26–27
Bowen, Will, 93, 94, 225
The Boys in Autumn (Sabath), 160
Bradbury, Malcolm, 180
Bradley, David, 110–12, 123, 191–93, *193*, 200,
202–203, 232, 248
Brazil, 144, 208
Breinig, Helmbrecht, 240
Brenner, Gerry, 242
Britain. *See* United Kingdom
Britton, Wesley, 238
Brock, Darryl, 158
Brown, Doc Buck, 45
Brown, F. X., 163
Brown University, 139, 237
Budd, Louis J., 81–82, 83, 131, 163, 199, 208, 233
Buell, Dwight, 172
Buffalo Express, 75, 81, 82–83, 169
Buffett, Jimmy, 128, 139, 238
Bugs Bunny (cartoon character), 147, 149–50
The Burden of Southern History (Woodward), 15
Burr, James, 53, 55–56, *56*, 214
Butcher, Philip, 103, 218, 228
Buultjens, Ralph, 185, 246

Cable, George Washington, 227
Cambridge, England, 183–85, 189–90
Carkeet, David, 155–57
Carlyle, Thomas, 78–79, 219
Carroll, Lewis, 152
Cartoons. *See* Animation
Castillo, Susan, 239
Catcher in the Rye (Salinger), 189
Cather, Willa, 191
"The Celebrated Jumping Frog of Calaveras
County" (Twain), 93, 138, 139, 238
Cemeteries: black cemeteries in St. Louis, 67;
Confederate soldiers buried in Elmira, 220;
"Injun Joe" headstone in Hannibal, 46–48, *47*
Century magazine, 40, 227
Chadwick Joshua, Jocelyn, 121, 232
Chaney, James, 5, 200
The Chaneysville Incident (Bradley), 110, 193, 248
Charles, Prince, 188–89
Charley (slave), 20, 22
Cheers, 10
Chesnutt, Charles, 110
Chinese in San Francisco, 82, 83, 97, 119, 146,
222
Chinese responses to Twain, 144, 145, 146, 207
Christian Science (Twain), 151
"The Chronicle of Young Satan" (Twain), 153
Churches: A.M.E. Church, Hannibal, 46; Baptist
church, 15, 38; Douglass A.M.E. Zion church,
Elmira, 81, 222; First Presbyterian Church,
Elmira, 81; Park Church (Independent
Presbyterian Church), Elmira, 81; Presbyterian

church, 67, 81, 219; Southern Baptist Convention, 37–38, 40, 71, 81, 212
Civil Rights Act of 1875, 198, 202
Civil Rights Movement, 4–5, 52, 53, 109, 166, 200
Civil War, 32, 40, 58, 80, 88, 116–17, 215
Clare College, 183–85
Clare Hall, 183, 185
Clark, Capt. William, 24–25
Claymation, 175
Clemens, Clara, 91, 138, 140, 155, 237
Clemens, Henry, 30, 179
Clemens, James Ross, 134
Clemens, Jean, 94, 140, 155
Clemens, John Marshall, 20, 53, 57, 75, 81
Clemens, Olivia (Livy) Langdon, 74, 75, 81, 85, 95, 140, 152, 155
Clemens, Orion, 53, 54, 57, 131, 170, 174–75
Clemens, Samuel Langhorne. *See* Twain, Mark
Clemens, Susy, 91, 140
Clinton, Bill, 63, 73
Cole, Marvin, 163
Coleman, Howard, 84
The Color Purple (Walker), 189
Computer technology, 167–69, 170, 172, 175, 177–78, 243–46
Condol, Leon W., 224
Condol, Louise Florence Washington, 90, 91, 224–25
A Connecticut Yankee in King Arthur's Court (Twain), 7, 14, 31, 64, 139, 173; dramatizations and adaptations of, 128, 139, 141–42, 144, 161–62, 238; influence of, on other writers' works, 154, 161–62, 191; slavery in, 22, 29, 59–60, 87–88; technology in, 179–80
Contests, 31, 33, 137–38, 161–62
Convict-lease system, 227
Conway, W. C., 48
Cooper, James Fenimore, 192, 195
Cope, Virginia, 246
Copley (James S.) Library, La Jolla, Calif., 176, 228, 237, 244
Cord, Mary Ann, 78, 85–90, *86*, 97, 98, 224
"Corn-Pone Opinions" (Twain), 230
Cornwallis, Lord, 185
Cowboys gang, 50
Coyote. *See* Wile E. Coyote (cartoon character)
Crane, Hiram, 80
Crane, Susan Langdon, 75, 81, 91, 220
Crane, Theodore, 80, 219
Crèvecoeur, J. Hector St. John de, 8
Crew, Spencer R., 216
Crow, Hiawatha, 37, 48, 51–53
"Cultural Landscape of the Plantation" (Vlach), 62
Cummings, Sherwood, 178, 244
Czech Republic/Czechoslovakia, 8, 189, 190

Daghlian, Carlos, 208, 239
Dan'l, Uncle (slave), 22
Dark Witness: When Black People Should Be Sacrificed (Again) (Wiley), 196

Daughters of the Confederacy, 220
Davis, Ernie, 90, 201
Davis, Gwen, 161
The Day They Came to Arrest the Book (Hentoff), 161
Death on the Mississippi (Heck), 158
The Deerslayer (Cooper), 195
DeVoto, Bernard, 119
"Disgraceful Persecution of a Boy" (Twain), 82, 83, 93
Disney, Walt, 147
Disneyland, 147, 181
Domestic Manners of the Americans (Trollope), 8
Douglas, Anna S., 48
Douglas, Joe, 42–48, 47, 213
Douglass, Frederick, 41, 52, 80, 81, 88–89, 94–97, 118, 178, 198, 213, 221, 222, 226–27
Douglass A.M.E. Zion church, 81, 222
Doyno, Victor, 198–99, 226, 227
Dramatizations and adaptations of Twain's works, 10, 92–93, 100, 105, 128, 138–44, 160, 237–39
Du Bois, W. E. B., 119, 122
Dunbar, Paul Laurence, 119
Duneka, Frederick A., 153
Dunning school of Southern history, 117, 200
Durczak, Jerzy, 207

Eaton, Winifred, 80
Edison, Thomas, 133, 139, 173
Education, 4–6, 8–11, 15–17, 33–37, 40, 41, 46, 49–51, 66–73, 81, 101–104, 107, 114–24, 133, 139, 145–46, 164, 185, 196–97, 202, 207–8, 212–13, 218, 229, 248–49. *See also* Brown University; Textbooks; University of Texas at Austin; West Point; Yale University
1876 (Vidal), 155
Eisenhower, Dwight, 166
Elijah Lovejoy Society, 67, 68
Eliot, T. S., 184
Ellison, Ralph, 51, 73, 110–13, 123, 184, 191, 193, *195*, 197–98, 200–202, 230
Elmira, N.Y., 7, 74–85, 77, 89–92, 94, 96, 97, 163, 219, 220, 223–25; African Americans in, 79–81, 84–91, 94–97, 201, 220–27. *See also* Quarry Farm
Elmira College, 74, 75, 226
Emerson, Ralph Waldo, 135, 136
England. *See* United Kingdom
English Society for Psychical Research, 152
Evers, Medgar, 109
Eve's Diary (Twain), 7, 144, 151
Extract from Captain Stormfield's Visit to Heaven (Twain), 154
"Extracts from Adam's Diary" (Twain), 7, 144

The Fabulous Riverboat (Farmer), 154–55
Facen, Rev. Ann, 37, 48–51
"Facts Concerning the Recent Carnival of Crime in Connecticut" (Twain), 93
Fagan, Robert T., 115
"Family Sketch" (Twain), 228
The Famished Road (Okri), 189–90

Farmer, Philip Jos, 154–55
Fast, Howard, 161
Fatout, Paul, 226
Faulkner, William, 184, 191
Fender, Stephen, 239
Fidler, Roger, 134
Films. *See* Movies
Finn, Huckleberry (fictional character), 4, 6, 21, 23, 26, 27, 28, 32–34, 39, 44, 51, 52, 58–64, 88, 98, 111–13, 137, 139, *140*, 151, 158–60, 185–89, 196, 202, 211, 230, 231. *See also Adventures of Huckleberry Finn* (Twain)
Finn, Pap (fictional character), 5–6, 109
First Presbyterian Church, 81
Fishkin, Joey, 162
Fitzgerald, F. Scott, 191
Five Famous Missourians (Williams), 25
Flipper, Henry O., 72
Florida, Mo., 21, 52, 91
Following the Equator (Twain), 14, 128, 139, 152–53
Foreign responses to Twain, 7, 8, 13, 26–27, 107, 144–46, 164, 183–85, 189–90, 207–8, 246–48
Forte, Dixie M., 37, 48, 49, 51
Foss, Lukas, 238
Frank, Michael, 152
Franklin, Benjamin, 135, 136
Free blacks in the slave South, 15–17, 28–29, 66, 208–9
French responses to Twain, 146
The French Revolution (Carlyle), 78–79, 219
The Further Adventures of Huckleberry Finn (Matthews), 159–60

Gabrilowitsch, Clara Clemens, 26, 27
Gabrilowitsch, Nina, 27
Galaxy Magazine, 82, 83
Garfield, James, 95, 96, 226
Garrison, William Lloyd, 80, 221
Gates, Henry Louis Jr., 107
Georgoudaki, Katia, 207–8
Gerhardt, Karl and/or Josephine, 58, 175, 176
German responses to Twain, 144, 146
"Ghost Story" (Twain), 168–69, 172
Gibson, William M., 153, 241
The Gilded Age (Twain and Warner), 7, 154, 179, 180
Gleason, Silas, 221
Glover, Samuel Taylor, 56
God Bless U, Daughter (Swanson), 152
Goldberg, Jim, 137
"The Golden Arm" (Twain), 139
Gone With the Wind (Mitchell), 92–93
Good Ol' Boys Roundup, 217
Goodman, Andrew, 5, 200
Grant, Ulysses S., 215
Great Britain. *See* United Kingdom
Great Twain Robbery: A Comedy Caper (Guntrum), 161
Greece, 207–8
Greener, Richard, 103

Gregory, Dick, 108–10
Griffin, George, 98–99, 227
Griggsby, Estel, 40–41
Guest, Edgar, 18
Guntrum, Robert, 161
Gutenberg, Johannes, 173, 181
"Gutenberg Project," 181

Hagood, Hurley, 25, 37, 46, 47, 49, 53–54, 213
Hagood, Roberta, 25, 37, 46, 49, 54, 213
Halley's Comet, 9, 26, 155
Hannibal (slave), 24–25
Hannibal, Mo., 7, 13–21, *19*, 24–54, 27, 56, 61–67, 91–92, 94, 137–38, 163–64, 201; African Americans in, 13, 14, 18, 19–25, 28–29, 34–37, 40–57, 61, 94, 201, 209, 210; tourism in, 17–19, *19*, *20*, 25–26, 27, 31–35, 39–40, 42–44, 47, 48, 61–66, 137–38
Hannibal Yesterdays (Hagood and Hagood), 46
Hannicks, John and Ellen, 29
Harper, Phillip, 249
Harte, Bret, 157
Hartford, Conn., 3–4, 7, 98–100, 108–10, 113–14
Hartford Courant, 57–58, 161–62, 173
Haufrecht, Herbert, 238
"Haunted House on Hill Street" Wax Museum (Hannibal), 18, *20*, 65
Hawthorne, Nathaniel, 152, 184
Hecht, Ben, 191
Heck, Peter J., 158
Heisel, Sharon, 161
Hemingway, Ernest, 7, 49, 112, 157, 184, 191, 198
Henry, William A. III, 160
Hentoff, Nat, 100, 161
Herbert, Frank, 154
Herbolsheimer, Bern, 238
Hess, Cliff, 139
Higginson, Thomas Wentworth, 144
Hirsch, E. D. Jr., 113
Hirst, Robert, 136
Hobby, W. P., 16, 200, 209
Hobby, William P. Jr. 16, 209
Hochman, Barbara, 209, 239
Holbrook, Hal, 113, 164–67, *168*, 230
Holcombe, R. I., 57, 214
Hollings, Ernest "Fritz," 73
Hornung, Alfred, 240
Horton, James O., 61–62, 216
"How I Edited an Agricultural Paper Once" (Twain), 4
Howells, William Dean, 72, 158, 174, 215
Hubbard, Elbert, 151
Huckleberry Fiend (Smith), 160–61
"Huckleberry Finn" (song), 139, *140*, 237
Huckleberry Finn (Twain). *See Adventures of Huckleberry Finn* (Twain); Finn, Huckleberry (fictional character)
Hughes, Langston, 51, 110, 111, 191
Hutchings, Emily Grant, 151–52, 153

I Been There Before (Carkeet), 155–57
"I Want to Know Why" (Anderson), 189

IBM VoiceType Dictation system, 167–69, 170, 172, 243
Ichikawa, Hiroyoshi, 207, 243
If I Never Get Back (Brock), 158
Impersonators, 9–10, 26, 102–103, 113, 128, 163–67, *168*, 178, 230, 242
Independent Presbyterian Church, Elmira, 81
Indian responses to Twain, 144, 146, 208
Injun Joe (fictional character), 42, 43–44, 45–48, 47
The Innocents Abroad (Twain), 17, 77, 118–20, 146, 179, 186, 187, 189
International Committee Against Racism, 37
Internet, 10, 180–81, 246
Inventions. *See* Technology
Invisible Man (Ellison), 193, 201
Irish responses to Twain, 144, 145
Is Shakespeare Dead? (Twain), 151
Israeli responses to Twain, 144, 207
Italian responses to Twain, 146

James, Henry, 183
"Jane Lampton Clemens" (Twain), 209
Jap Herron (Hutchings), 151–52, 153
Japanese responses to Twain, 7, 8, 13, 144, 145, 189, 190, 208
Jefferson, Thomas, 24, 88
Jerome, Helen, 89–90, 220
Jerome, Robert, 89–90
Jerry (slave), 22, 51, 111, 230
Jesch, Gary, 178
Jim (fictional character), 5, 20, 21, 39, 49, 52, 58–59, 63, 65, 66, 88, 115, 119–21, 137, 145, 159, 160, 196–99, 231
Joan of Arc (Twain), 156–57
Johnson, W. F., 43
Jones, A. W., 98
Jones, Chuck, 147–50
Jones, John W., 80–81, 84, 95, 220
Jones, LeRoi, 192

Kamei, Shunsuki, 239
Kaplan, Justin, 162, 164, 225
Kar, Prafulla C., 145–46, 208, 240
Karsavina, Jean, 238
Kennedy, John F., 63
Kern, Jerome, 238
A Kid in King Arthur's Court, 142
The Kid (Seelye), 242
Kinch, J. C. B., 240
Kipling, Rudyard, 146, 235
Kiskis, Michael, 75, 226
KKK. *See* Ku Klux Klan
Knight, Emma, 19, 21, 22, 210
Knights of the White Camellia, 118
Koppel, Ted, 100, 228
Kruse, Horst H., 240
Ku Klux Klan (KKK), 36–37, 38, 41, 51–52, 68–69, 118, 166, 218, 231
Kutrieh, A. R., 208

Lang, Andrew, 185
Langdon, Charles, 74–75, 85, 97, 221
Langdon, Ida, 97
Langdon, Jervis, 74–75, *76*, 79–81, 85, 95, 219–21
Langdon, Jervis, Jr., 75
Langdon, Olivia. *See* Clemens, Olivia Langdon
Langdon, Olivia Lewis, 81, 84, 221, 226
Langdon, Susan. *See* Crane, Susan Langdon
Lawrence, Joey, 141
Layne, McAvoy, 164, 178
Le Guin, Ursula, 191, 248
Lennon, John A., 54
Lester, Julius, 184
Letters from an American Farmer (Crèvecoeur), 8
Levi, Paul Alan, 238
Lewis, John T., 91, 95, 96, 226–27
Lewis, Sam M., 139
Lewis, Susy, 91
Library of Congress, 62
"Lie of silent assertion," 60–61, 201
Life and Times (Douglass), 88–89
Life magazine, 33
Life on the Mississippi (Twain), 14, 28, 29, 34, 39, 61, 65, 69, 97, 124–25, 154, 175, 191
Lincoln, Abraham, 73, 95, 136
Lincoln, Robert T., 73
Lincoln University, 98, 99, 103, 104, 228, 229
The Literary Reputation of Mark Twain (Asselineau), 146
Littlefield, George W., 116
Lobato, Monteiro, 144
London *Times*, 135
Lynchings. *See* Racism
Lyon, Isabel, 165
Lystra, Karen, 158

Mahan, George A., 26, 27
Malone, Cheryl, 116
"The Man That Corrupted Hadleyburg" (Twain), 4, 14, 144, 161
Man with a Million, 141
Maoris, 117
"Marching Through Georgia" (Work), 57–58, 215
Marion College (Eels College), 55
"Mark Me Twain," 238
"Mark Twain: Portrait for Orchestra," 238
Mark Twain Book and Gift Shop (Hannibal), 31–33
Mark Twain Boyhood Home and Museum (Hannibal), 17–18, 26, 27, 29–31, 39, 239
Mark Twain Bridge (Hannibal), 26, 63, 64
Mark Twain Cave (Hannibal), 42
Mark Twain Centennial (1935), 26, 27
Mark Twain Circle of America, 63, 165
Mark Twain Day, 26
Mark Twain Forum (e-mail discussion group), 181, 246
Mark Twain Historic District (Hannibal), 17–18, 24, 40, 41, 65–66
Mark Twain House (Hartford), 3–4, 98–100, *99*, 108–10, 113–14

Mark Twain Lighthouse (Hannibal), 26, 32, 63
Mark Twain Murders (Skom), 161
Mark Twain obituary class, 135
Mark Twain Outdoor Theater, 39–40, 64
Mark Twain Papers/Project (Berkeley), 103, 111, 136, 152, 156–57, 228, 229, 244
"Mark Twain Resources on the World Wide Web" (Zwick), 181
Mark Twain riverboat (Hannibal), 62–65
Mark Twain Sesquicentennial (1985), 26
Mark Twain steamboat (Disneyland), 147, 181
"Mark Twain Suite," 238
Mark Twain/Halley's Comet U.S. Aerogramme, 26
"Mark Twain's First One Hundred Years," 26
Marsh, Tyrone, 221
Marshall, Thurgood, 105, 106
Marszalek, John, 218
Martin Luther King Day, 51
Mason, Bobbie Ann, 180, 188, *193*
Materassi, Mario, 146, 240
Matory, Bill, 121
Matthews, Greg, 159–60
Matthews, Nancy, 137
McComb, Gordon, 175
McDowell, Edwin, 106
McFeeley, William, 88, 89
McGuinn, Warner T., 99–107, 119, 228–29
Meachum, John Berry, 15–17, 24, 40, 67, 68, 200, 208–9, 212
Media interpretations. *See* Movies; Television
Meredith, James, 166
Mergenthaler, Ottmar, 169, 170, 173
Messent, Peter, 185, 247
Messerle, Brad, 167–70, 172
Michaelson, Bruce, 247
Miller, Arthur, 191
Miller, Henry, 191
Miller, James A., 237
Miller, Roger, 143
A Million to Juan, 141, *142*
Mission Institute, 53, 55
Mississippi, 34, 83, 218
Mississippi: An American Journey (Walton), 83, 222
Mississippi River, 24, 25, 27, 28, 62–64, 69, 124–25
Missouri. *See* Florida, Mo.; Hannibal, Mo.; St. Louis, Mo.
Mitchell, Kirk, 157
Montgomery, Yvonne, 160
More Adventures of Huckleberry Finn (Wood), 159
Morgan, Hank (fictional character), 29, 87–88, 180
Morris, Willie, 191
Morrison, Toni, 111, 191, *194*, 198, 200, 202
Morse, Samuel F. B., 173
Mosley, Walter, 128
The Most Remarkable Echo in the World (Partridge and Partridge), 152–53
Movies, 10, 92–93, 128, 138–44

Muller-Van Santen, Julia J. G., 239
Murder Stalks the Circle (Thayer), 160
Murray, Albert, 189
Murrell's Gang, 46
Museums, 34, 51, 62, 114, 216
Music, 39, 57–58, 95, 122, 139, *140*, 143, 144, 215, 226, 237, 238
"My First Lie and How I Got Out of It" (Twain), 14
My So-Called Life, 143
"Mysterious Stranger" stories (Twain), 14, 154
Mystery novels, 160–61

NAACP, 16, 34, 49, 50, 90, 104
Nagawara, Makoto, 145, 208, 239, 240
Napier, Susan J., 247
Nathan, John, 190, 247
National Park Service, 61, 216
National Underground Railroad Freedom Center, 62, 217
Neider, Charles, 153–154
Nelson, David, 55
Netherlands, 145
Never the Twain (Mitchell), 157
New York Times, 66, 98, 101, 106, 111, 180
New York University, 197, 254
New Zealand, 117
Nigeria, 8, 189–90
Nigger (Gregory), 108
"Nile: Passage to Egypt," 175
Notoji, Masako, 13
Nye, Bill, 133, 135

Oates, Joyce Carol, 155
Oē, Kenzaburo, 190, *194*
Okri, Ben, 189–90, *194*
Okubo, Hiroshi, 7, 207
"Old Times on the Mississippi" (Twain), 92
Olsen, Tillie, 191
"On Foreign Critics" (Twain), 23, 211
"£1,000,000 Bank Note" (Twain), 139, 141, *142*, 144
"One-drop rule," 43, 72, 113
"Only a Nigger" (Twain), 82, 83, 93, 97, 108
Operas, 139, 237–38
Oral History (Smith), 189
Our Mark Twain: The Making of His Public Personality (Budd), 131
Oxford English Dictionary, 189
Oxford Mark Twain, 10–11
Oxford University, 185
Ozick, Cynthia, 180

Page, Clarence, 108
Paige, James William, 169–70, 172–73
Paige typesetter, 169–70, *171*, 172–73, 179
Paine, Albert Bigelow, 49, 82, 83, 153, 219
Park Church (Independent Presbyterian Church), Elmira, 81, 221
Partridge, D. C. and H. M., 152–53
Patell, Cyrus, 248
Peck, Gregory, 141

Peck, John Mason, 15
Perry, Tom, 231
Petzold, Charles, 175
Philadelphia Anti-Slavery Society, 80
Philippines, 35
Pickle, David J., 16, 200, 209
Pickle, Jake, 16, 209
Plana, Tony, *142*
Plessy v Ferguson, 35
Poe, Edgar Allan, 152–53
Pohl, Frederik, 190–91, *193*, 248
Poland, 207
Poor Richard's Almanac (Franklin), 135
Pop, Vasile, 151, 241
Porter, Charles Ethan, 98
Portugal, 145
Powers, Ron, 24–25
Presbyterian church, 67, 81, 218
The Prince and the Pauper (Twain), 31, 139–41, 144
Prince for a Day, 141
Princess and the Pauper: An Erotic Fairy Tale (Davis), 161
Pritchard, Lee H., 238
"The Private History of a Campaign That Failed" (Twain), 14
Pudd'nhead Wilson (Twain), 14, 21, 56, 72, 90, 135, 197, 202, 248–49
Pulliam, Keshia Knight, 142

Quaker City tour, 74, 186, 225
Quarry Farm, 75, 77–78, *77*, 85, 90–91, 95, 219, 223–26
Quotations and misattributed quotes of Twain, 10, 49, 133–37

Race relations: Civil Rights Movement, 4–5, 52, 53, 109, 166, 200; education on, in Hannibal, 35–37; in Hannibal, 35–37, 40–41, 49, 50, 51; Huck's friendship with Jim in *Huckleberry Finn*, 52, 58–59, 63, 65; interracial friendships in Hannibal, 36, 41; interracial friendships of Twain's children, 91, 224; in post-Reconstruction era, 96–97, 105–106, 118, 119, 197–203, 227, 231; and Southern Baptist Convention, 37–38, 212; in *Tom Sawyer*, 225; Twain's focus on, 9; Twain's interracial boyhood friendships, 91. *See also* African-American history and culture; Racism; Underground Railroad
Rachel, Aunt (fictional character), 85, 87
Racism: in Elmira, 89–90; in Greenwich, Conn., 66–67; in Hannibal, 36–37, 40–41, 49–52; Malcolm X on, 6; in New Zealand, 117; "nigger" as term, 49, 67, 100, 107, 108, 109, 110, 115, 116, 197, 217; in post-Reconstruction era, 96–97, 105–106, 118, 119, 197–203, 227, 231; stereotypes, 88, 92–93; in Texas, 16, 117; of Truman, 52–53; of Twain, 73–74; Twain's antiracism, 4–6, 9, 41, 51, 59–61, 71–72, 74, 82–84, 90, 93, 94, 97, 100–101, 106–109, 113–14, 119–21, 123, 197, 199–202, 222, 227; violence

against blacks, 4–5, 14, 21–22, 35, 72–73, 82, 82–83, 109, 114, 118, 200, 202, 210, 222, 231; at West Point, 71–73, 199. *See also Adventures of Huckleberry Finn* (Twain); Convict-lease system; Cowboys gang; Ku Klux Klan; *Pudd'nhead Wilson* (Twain); Race relations; Slaves and slavery; Southern Baptist Convention
Raft on the River, 139
Raja aur Runk, 144
Rampersad, Arnold, 230
Rasmussen, Kent, 140, 238
Reagan, Ronald, 26
Reda, Linda, 243
"Reflections of Mark Twain," 39–40, 64
"The Remittance Man" (Buffett), 139
Renfro, Brad, 143
Renyi, Mei, 207, 239
"Return of Mark Twain" (Winkler), 152
Rice, Patricia, 37
Richler, Mordecai, 188–89
Richters, Ken, 163–64
Rifle Clubs of South Carolina, 118
Rigaud, Edwin J., 217
Road Runner cartoons, 146–50
RoBards, Archibald, 54
Roberts, Taylor, 246, 247
Robinson, Forrest, 60, 216
Robotics, 177–78
Rodriguez, Paul, 141, *142*
Rogers, Will, 142
Roosevelt, Franklin Delano, 7, 26
Roosevelt, Theodore, 66
Rosario, Bert, *142*
Rosarossa, Maria Alejandra, 185, 207, 239, 246
Rose, John C., 105–106
Rosenblatt, Roger, 158
Rosenbloom, Julia, 196–97, 248
Roughing It (Twain), 139, 148–49, 157
Rule of the Bone (Banks), 189
Russell, Walter, 27
Russia, 107, 144

Sabath, Bernard, 160
St. Louis, Mo., 15–17, 26, 67–69
Salinger, J. D., 189
Santirocco, Matthew S., 197, 248–49
Saroyan, William, 191
Sasaki, Kuni, 144
Saudi Arabia, 208
Sawyer, Tom (fictional character), 4, 5, 17, 26, 27, 31, 32, 34, 44, 65–66, 119–21, 137–38, 185, 198. *See also The Adventures of Tom Sawyer* (Twain); Tom Sawyer Days (Hannibal)
Scavengers (Montgomery), 160
Scharnhorst, Gary, 92
Schools. *See* Education
Schwartz, Richard, 35–37, 41, 48, 49
Schwerner, Michael, 5, 200
Scicchitano, David A., 249
Science. *See* Technology
Science fiction, 10, 150–51, 154–55, 161, 190–91

"A Scrap of Curious History" (Twain), 14
Seelye, John, 159, 160, 241, 242
Segregation: in Hannibal, 36, 40–41; in post-Reconstruction era, 96–97, 105–106, 118, 119, 227. *See also* Race relations; Racism
Selznick, David O., 92, 143, 239
Shakespeare, William, 31, 78, 139, 241
Shallat, Ben, 238
Sharlow, Gretchen, 77, 89, 220, 238
Shaw, George Bernard, 7
Shermeister, Pam, 248
Shillady, John R., 16, 209
Siberia, 189
Siebert, William, 80, 220
Silas Timberman (Fast), 161
Skandera-Trombley, Laura, 221
Sketches, Old and New (Twain), 90, 169
Skom, Edith, 161
Skvorecky, Josef, 190
Slave narratives, 110
Slaves and slavery: abolitionism, 53, 55, 74, 75, 79–81, 95, 197, 208, 221; advertisements for sale of slaves, 18, 209; Douglass on, 88–89; education of, 15–17, 208–9; exhibits on, 62, 68; in Hannibal, Mo., 13, 14, 18, 19–24, 51, 209–10; invisibility of, 39–40; liberation of, in U.S. and England, 73, 95; in movies, 92–93; murders of slaves, 14, 21–22, 210; oral tradition of slaves, 22, 51; runaway slaves, 21–22, 35, 53–57, 63, 80, 84–85, 210, 223; separation of families through sale of slaves, 85, 87–88; slave trade, 18, 20, 22, 68, 85, 87–88, 209–10; Southern Baptist Convention resolution on, 37–38, 40, 212; textbook portrayal of, 122, 123; Twain's comments on, 23, 41, 51, 53–54, 59–61, 85, 87–88, 89; in Twain's works, 20, 21, 22, 29, 59–60, 87–88, 89, 92–94; and Underground Railroad, 41, 49, 53–57, 62, 67, 79–80, 84–85, 201, 216–17, 220–21, 223. *See also* Black John; Blandford, Alsace; Blandford, Thomas; Charley; Cord, Mary Ann; Dan'l, Uncle; Hannibal; Jerry; Knight, Emma; Meachum, John Berry; Smith, Clay; Turner, Nat
Smiley, Jane, 195–97, 202
Smith, Clay, 19–20, 21, 22, 210
Smith, Gerrit, 80, 221
Smith, Henry Nash, 13, 92, 199, 208
Smith, Julie, 160–61
Smith, Lee, 189
Smith, Nathan, 80
Sommers, Stephen, 143
Songs. *See* Music
Soren, Gretchen Sullivan, 220
Southern Baptist Convention, 37–38, 40, 71, 81, 212
Soviet Union, 144, 164, 189
Spencer, Elizabeth, 191
Spiritualism, 151–52
Sri Lanka, 185
Star Trek: The Next Generation, 10, 150–51
Statues, 26, 68, 123
Steamboats, 28–29, 62–65, 147, 158, 179

Stein, Gertrude, 191
Steinbrink, Jeffrey, 83, 225
Stewart, James T., 191–92
Still, William, 80, 84
Stiner, Nancy, 101–103
Stiner, Richard, 102
Stoneley, Peter, 145, 146, 185, 207, 239, 240, 247
Storytelling, 22, 29, 51, 85, 87
Stowe, Harriet Beecher, 196, 197, 202–203
Strickland, Agnes, 78
Supreme Court, 198
Swanson, Mildred Burris, 152
Sweets, Henry III, 30–31, 163
Syracuse University, 89

Tabscott, Robert, 67–69, 201, 209, 217–218
Tang–Fengyu. *See* Xun, Lu
Tanner, Tony, 185, 199, 247
Taylor, Meshach, 100
Technology, 169–81; in *The American Claimant*, 180; animatronics and robotics, 177–78; in Appel's *Twice Upon a Time*, 157; computer technology, 167–69, 170, 172, 175, 177–78, 243–46; in *A Connecticut Yankee in King Arthur's Court*, 179–80; and *A Connecticut Yankee in King Arthur's Court* adaptations, 128, 141–42; IBM VoiceType Dictation system, 167–69, 170, 172, 243; Paige typesetter, 169–70, 171, 172–73, 179; telephone, 4, 173, 179–80; TWAIN application programming interface, 175, 177, 244–45; Twain's inventions, 10, 78, 173; typewriter, 174–76; and World War I, 180
Telephone, 4, 173, 179–80
Television, 10, 100, 107–108, 134, 141–44, 150–51, 164, 180, 218
Tesla, Nikola, 173, 177, 244
Textbooks, 35–36, 41, 114–16, 122, 123. *See also* Education
"That's What Living Means to Me" (Buffett), 139
Thayer, Lee, 160
Theater productions, 39–40, 100, 105, 138–44, 160
Thomason, Jerry, 179
Thompson, George, 53, 55–56, 56, 214–15
Thurmond, Strom, 73
Thurston, Ariel, 80–81, 221
Time magazine, 158
Tom and Huck, 143
Tom Sawyer (Twain). *See Adventures of Tom Sawyer* (Twain)
Tom Sawyer Abroad (Twain), 154
Tom Sawyer Days (Hannibal), 17, 31, 33, 137–38
Tom Sawyer Grows Up (Wood), 159
"Tom Sawyer's Conspiracy" (Twain), 14
Tom Sawyer's Island (Disneyland), 147
Tourism and travel: Twain on, 9, 17, 77; Twain's worldwide travel, 7–8, 14, 21, 74–75, 146, 186. *See also* Hannibal, Mo.
Townsend, Charles, 184–85
Train Whistle Guitar (Murray), 189
A Tramp Abroad (Twain), 154, 191
Trinity College, 189–90

Trollope, Frances, 8
The true adventures of Huckleberry Finn (Seelye), 159
"A True Story, Repeated Word for Word as I Heard It" (Twain), 85, 87, 90, 98
Truman, Harry, 52–53
Turner, Frederick Jackson, 40
Turner, James Henry, 67
Turner, James Milton, 40, 212
Turner, Nat, 40
Tuskegee Airmen, 73
Twain impersonators. *See* Impersonators
Twain, Mark. *See also* Elmira, N.Y.; Hannibal, Mo.; Hartford, Conn.; headings beginning with Mark Twain; and specific titles of works
—attitudes and beliefs: antiracism, 4–6, 9, 41, 51, 59–61, 71–72, 74, 82–84, 90, 93, 94, 97, 100–101, 106–109, 113–14, 119–21, 123, 197, 199–202, 200, 222; on civilization, 23; on copyright, 181; on "Americanness," 7–8, 63, 69, 144–45; racist attitudes of, 73–74; on slavery, 14, 20–24, 51, 53–54, 59–61, 74–89, 92–94, 201; on technology, 10, 169–70, 172–81, 244–45
—biographical information: boyhood, 18, 19–24, 30, 51, 53–54, 91; business investments, 172–74, 179; children, 26, 27, 91, 94, 140, 224; courtship, 74–75; death, 25, 32; fame and popularity, 131, 144–45; financial support for blacks, 98, 101–106, 119; images projected by, 11, 113–14, 127–28; inventions, 10, 78, 173; personal library and reading, 58, 78–79, 88, 219, 224; worldwide travels, 7–8, 14, 21, 74–75, 146, 186
—influence on other writers, 8, 51, 110–13, 188–96
—in popular culture: advertising, 9, 128–31, *130*, *132*; businesses and places named after Twain and Twain's fictional characters, 9, 17, *19*, 26, 38, 48, 53, 89, 131, *133*; dramatizations and adaptations of Twain's works, 137–50; manuscripts and plots of Twain in other authors' books, 160–63; poetry about Twain, 18; "posthumous" writing by Twain, 151–54; sketch of Twain by Chuck Jones, *148*; Twain as android, 177–78; Twain as character in others' works, 10, 150–51, 154–58
—significance and contributions of, 6–11, 73, 183–203
—translations of works, 7, 11, 144
"Twain's World," 175
Twice upon a Time (Appel), 157–58
Typewriter, 174–76

Uncle Tom's Cabin (Stowe), 196, 197, 202–203
The Underground Rail Road (Still), 84
Underground Railroad, 41, 49, 53–57, 62, 67, 79–80, 84–85, 201, 216–17, 220–21, 223. *See also* Abolitionism; Slaves and slavery
Underground Railroad Freedom Center, 62, 217
Unidentified Flying Oddball, 128, 142
United Kingdom, 145, 146, 183–85, 189–90, 207
"The United State of Lyncherdom" (Twain), 114

University of Cambridge, 183–85, 189–90
University of Maryland, 90
University of Texas at Austin, 16, 116–25

Vidal, Gore, 155
Vietnam War, 166
"Villagers of 1840–3" (Twain), 29
Vinton, Will, 175
Virginia, 218
"Virtual Mark Twain," 178
Vlach, John, 62
VoiceType Dictation system. *See* IBM VoiceType Dictation system

Wagner, Richard, 133
Walker, Alice, 189
Wallace, John, 100, 107–108, 159, 228
Walt Disney World, 177–78
Walton, Anthony, 83, 222
"The War Prayer" (Twain), 139, 238
"War Prayer Oratorio," 238
Warner, Charles Dudley, 137, 154, 179
Warner Brothers, 147
Was Huck Black? (Fishkin), 109–13, 158
Washington, Henry Crummell, 85, 90, 223–24
Watt, James, 173
Wavada, Mike, 162, 227
Wayland, Francis, 103, 104
Webster, Charles, 173–74
Webster University, 67
Weill, Kurt, 139, 237
Wells, Ida B., 35, 114
West, Royce, 114–16
West Point, 71–73, 199, 218
What Black People Should Do Now (Wiley), 196
"What Have the Police Been Doing?" (Twain), 222
What Is Man? (Twain), 154, 155
White, Frank Marshall, 134
White Town Drowsing (Powers), 24–25
White Brotherhood, 118
Whittaker, Johnson C., 71–73, 74, 199, 218
Why Black People Tend to Shout (Wiley), 196
Wile E. Coyote (cartoon character), 147, 149
Wiley, Ralph, 196
Williams, Walter, 25
Williams, William Carlos, 184
Wilson, Bruce, 16, 209
Wilson, "Pudd'nhead" (fictional character), 56, 135, 202
Winkler, Eunice, 152
Wisbey, Herbert A., 223
"Wishbone" television version of *Tom Sawyer*, 142–43
Wood, Clement, 159
Wood, Grant, 187–88
Woodhouse, Mark, 79
Woodward, C. Vann, 15, *195*
WordPerfect 5.1: Macros and Templates (McComb), 175
Work, Alanson, 53, 55–57, *56*, 58, 214, 215
Work, Henry Clay, 57–58, 215

Wouk, Herman, 191
Wrapped in a Riddle (Heisel), 161
Wright, Richard, 90, 111, 191

X, Malcolm, 6, 116, 200
Xun, Lu, 144, 243
Xu-yi, Liu, 239, 240

Yale University, 100, 102, 103–104, 106
Yarbrough, Eugene W., 47
Yevtushenko, Yevgeny, 190, 247
Young, Joe, 139

Zinsser, William, 26
Zwick, Jim, 181, 246